Mathematics for Technicians 2

Mathematics for Technicians 2

D J Hancox

Head of Mathematics Department,
Coventry Technical College

GRANADA
London Toronto Sydney New York

Granada Publishing Limited – Technical Books Division
Frogmore, St Albans, Herts AL2 2NF
and
36 Golden Square, London W1R 4AH
866 United Nations Plaza, New York, NY 10017, USA
117 York Street, Sydney, NSW 2000, Australia
100 Skyway Avenue, Rexdale, Ontario, Canada M9W 3A6
61 Beach Road, Auckland, New Zealand

First published in Great Britain 1982 by Granada Publishing

British Library Cataloguing in Publication Data
    Hancox, D. J.
    Mathematics for technicians 2.
    1. Shop mathematics
    I. Title
    510.246   TJ165

ISBN 0-246-11725-7

Typeset by The Castlefield Press of Northampton
and printed in Great Britain by William Clowes (Beccles) Limited, Beccles and London

Granada ®
Granada Publishing ®

Preface

This book is the second in a series of books designed to provide the mathematics required for the TEC certificate courses. It covers the whole of the level 2 mathematics modules. The standard units were modified in 1980. This book covers the work of level 2 standard units and other work such as vectors, matrices, AP's, GP's and continued fractions which some colleges may have in their own units. The presentation has been designed to create interest in the subject, using practical examples from other subject areas wherever possible. In most cases each chapter represents a module from the TEC booklet. The student should study the text and worked examples in each section and then work all of the examples in the exercise. The above average student will find further examples in the 'General Examples' sections at the ends of chapters. Each chapter has a summary and self assessment test. The book is divided into three parts. The first part all level 2 students will follow. The second part is for students following a less academic route, this will also be useful to academic students who may find some of this work has been omitted from their programme. The third part is for those following an academic route, but much of this work the other students will meet in later years.

Each section is followed by the appropriate phase test. There is an examination at the end of each half unit. Full solutions and marking schemes are given to all self assessments, phase tests and examinations.

This book should be valuable to students in colleges or studying a similar course in school. The book has been written so that, if necessary, a student can work through the units with a minimum amount of teacher assistance.

I would like to thank the publishers for their helpful guidance and valuable suggestions, my daughter Jane for typing the manuscript and Les Burrows and Phillip Doughty for preparing the diagrams.

David J Hancox

Contents

1 The Calculator

It is assumed that all students will have a calculator with the following functions and one memory.

Addition	$+$
Subtraction	$-$
Multiplication	$\times$
Division	$\div$
Square root	$\sqrt{\ }$
Reciprocal	$\dfrac{1}{x}$
The sine	sin
The cosine	cos
The tangent	tan
The angle whose sine is	$\sin^{-1}$ or arc sin
The angle whose cosine is	$\cos^{-1}$ or arc cos
The angle whose tangent is	$\tan^{-1}$ or arc tan
The exponential	e^x
The natural logarithm	ln or $\log_e$
The power of X	X^Y
The value of π	π
Add to memory	M+
Subtract from memory	M—
Recall memory	RM
Clear memory	CM

Since there are so many different calculators available the actual details for the working of the calculator will not be given here. Students should check their calculators to see that they can obtain the correct answers to calculations. It is useful to work out the following calculations as a check. It is also worth noting at this stage that
(i) it is not always necessary to press the $=$ key in the middle of a calculation, e.g. $\frac{36}{9} + 5$ is obtained by $36 \div 9 + 5 =$, there is no need to press $=$ after the 9;
(ii) it is not always necessary to use X^Y when Y is small and an even whole number e.g. 8^2 is obtained by $8 \times =$ and 8^4 by $8 \times = \times =$ etc.

Calculator check

$39.87 + 87.54 + 46.26 = 173.67$

$76.43 - 19.29 = 57.14$

$97.84 - 17.65 + 26.81 = 107.00$

$8.74 \times 2.52 = 22.0248$

$82.36 \times 1.94 + 179.63 = 339.4084$

$(56.84 + 41.27) \times 75.34 = 7391.6074$

$1.85 \times 3.91 \times 12.23 = 88.465\,705$

$\dfrac{81.67}{1.83} = 44.628\,415$

$\dfrac{17.84 \times 1.97}{3.51} = 10.012\,763$

$\dfrac{1.84}{3.19} + \dfrac{12.15}{5.91} = 2.632\,640\,0$

$\dfrac{17.19}{5.23} - \dfrac{8.11}{4.83} = 1.607\,717\,8$

$\dfrac{1}{27.94} = 0.035\,790\,9$

$\sqrt{173.8} = 13.183\,322$

$5.81^{3.2} = 278.845\,7$

$e^{1.97} = 7.170\,671$

$e^{-2.57} = 0.076\,536$

$\ln 17.56 = 2.865\,623$

$\sin 37.8^\circ = 0.612\,907$

$\cos 101.5^\circ = -0.199\,368$

$\tan 75.6^\circ = 3.894\,741$

$\sin x = 0.762\,1, \ x = 49.649\,67^\circ$

$\cos x = 0.571\,3, \ x = 55.159\,06^\circ$

$\tan x = 1.743, \quad x = 60.156\,09^\circ$

1.1 CALCULATIONS

When a calculation is to be performed once only it is not worth spending time trying to find the quickest method. It is usually just sufficient to obtain the correct answer.

Example 1.1 Calculate the volume of a cylinder of diameter 67 mm and height 186 mm using the formula $V = \dfrac{\pi d^2 h}{4}$.

Also calculate the surface area using the formula
$S = 2\pi r(r + h)$

$d \times = \div 4 \times \pi \times h =$

The d^2 is worked out first to avoid entering d twice. The $\div 4$ is done early in the calculation to reduce the size of the number displayed. If all the multiplications in a calculation are performed first it may produce a number too large for the calculator.

$$67 \times = \div 4 \times \pi \times 186 = 655\,771 \text{ mm}^3$$

The volume of the cylinder is $655\,771 \text{ mm}^3$
To calculate the surface area:

$$\text{CM } d \div 2 = \text{M+} + h \times 2 \times \pi \times \text{RM} =$$

Before starting a calculation remember to clear the memory. It is usually necessary to press = after dividing by 2 before pressing M+, otherwise 2 will be entered into the memory

$$\text{CM } 67 \div 2 = \text{M+} + 186 \times 2 \times \pi \times \text{RM} = 46\,202$$

It can save time and it can be more accurate to programme a calculation that is to be repeated. That is form a simple flow chart.

Example 1.2 Use the formula $V = \frac{4}{3}\pi R^3$ to find the volumes of spheres with radii 21.5, 23, 24.5, 26, 27.5 and 29, all in mm.
The $\frac{4\pi}{3}$ is a constant in this calculation and should be worked out first.

$$4 \div 3 \times \pi = 4.188\,79$$

This constant is then stored in the memory and multiplied by the radius cubed for each volume.
Hence the flow chart is

$$\text{CM } 4 \div 3 \times \pi = \text{M+} \underline{R \times = \times R \times \text{RM} =}$$

The $R \times = \times R$ can be replaced by $R \times {}^Y3 =$

Radius (mm)	21.5	23	24.5	26	27.5	29
Volume (mm³)	41 630	50 965	61 601	73 622	87 114	102 160

Example 1.3 The time of swing of a simple pendulum is given by $t = 2\pi\sqrt{\dfrac{l}{g}}$ where $g = 9.81 \text{ ms}^{-2}$. Calculate the times of swing for the following lengths of pendulum, all given in metres: 0.6, 0.8, 1.0, 1.2, 1.4, 1.6, 1.8 and 2.

$$t = 2\pi\sqrt{\frac{l}{g}} = 2 \times \pi \times \frac{\sqrt{l}}{\sqrt{g}} = \frac{2 \times \pi}{\sqrt{g}} \times \sqrt{l}$$

The 2, π and g are all constant for this calculation. This constant should be worked out first.

$$\text{CM } 9.81\sqrt{} \text{ M+} 2 \times \pi \div \text{RM} = 2.006$$

This can be stored in the memory and the rest of the calculation is a square root and multiplication. The full calculation is:

$$\text{CM } 9.81\sqrt{} \text{ M+} 2 \times \pi \div \text{RM} = \text{CM M+} \underline{l \sqrt{} \times \text{RM} =}$$

Length (m)	0.6	0.8	1.0	1.2	1.4	1.6	1.8	2.0
Time of swing(s)	1.554	1.794	2.006	2.198	2.374	2.537	2.691	2.837

EXERCISE 1.1

1. Use the formula $A = \pi r^2$ to calculate the area of a circle of radius 87.9 mm.

2. Find the volume of a cone of height 1.74 m and base radius 0.54 m given that $V = \frac{1}{3}\pi r^2 h$

3. If $Z = \sqrt{\dfrac{x}{x + y}}$ find the value of Z when $x = 71.4$ and $y = 31.8$

4. Use the formula $I = \dfrac{nE}{R + nr}$ to calculate the value of I when $n = 7, E = 21.4, R = 87.6$ and $r = 3.44$

5. If $x = \dfrac{1 - t^2}{1 + t^2}$ calculate x when $t = 2.874$

In questions 6 to 10 use the formula given to complete the tables.

6. $v^2 = 2gh, g = 9.81 \text{ ms}^{-2}$

h (m)	2.87	4.76	7.93	12.84	27.13	34.76
v (ms⁻¹)						

7. $w = K\sqrt{d}, K = 1.947$

d	0.54	0.87	1.19	1.87	2.93	6.78
w						

8. $n = \dfrac{3t + 2}{t + 1}$

t	1.09	3.94	4.27	5.18	6.19	7.54
n						

9. $A = 2\pi r(r + h)$

r (m)	1.93	1.93	2.84	2.84	4.09	4.09
h (m)	3.76	5.61	3.76	5.61	3.76	5.61
A (m²)						

10. $Y = a + b\sqrt{x}, a = 19.72, b = 14.16$

x	217.8	154.3	116.9	98.72	73.8
y					

1.2 TRANSPOSITION OF FORMULAE AND CHECKING CALCULATIONS

In any formula the left hand side (LHS) is always equal to the right hand side (RHS). Hence whatever is done to the LHS must be done to the RHS.

(i) A term can be added to or subtracted from both sides

$$Y + 2 = X \Rightarrow Y = X - 2$$

(ii) The whole of both sides can be multiplied or divided by the same quantity (not zero)

$$3V = \pi r^2 h \Rightarrow V = \frac{\pi r^2 h}{3}$$

(iii) The whole of both sides can be raised to the same power (positive, negative or fractional)

$$v^2 = 2gh \Rightarrow v = \sqrt{2gh}$$

Example 1.4. Transpose the formula $A = \pi r^2$ to make r the subject.

$A = \pi r^2$ divide each side by π

$\dfrac{A}{\pi} = r^2$ take the square root of each side

$\sqrt{\dfrac{A}{\pi}} = r$ hence $r = \sqrt{\dfrac{A}{\pi}}$

Example 1.5. Transpose $V = \dfrac{2R}{R - r}$ to make R the subject.

$V = \dfrac{2R}{R - r}$ multiply both sides by $R - r$

$V(R - r) = 2R$ remove brackets

$VR - Vr = 2R$ add Vr to both sides

$VR = 2R + Vr$ subtract $2R$ from both sides

$VR - 2R = Vr$ collect into brackets

$R(V - 2) = Vr$ divide both sides by $V - 2$

$R = \dfrac{Vr}{V - 2}$

Example 1.6. Transpose $\dfrac{D}{d} = \sqrt{\dfrac{f + p}{f - p}}$ to make f the subject.

$\dfrac{D}{d} = \sqrt{\dfrac{f + p}{f - p}}$ square both sides

$\dfrac{D^2}{d^2} = \dfrac{f + p}{f - p}$ multiply both sides by d^2 and by $f - p$

$D^2(f - p) = d^2(f + p)$ multiply out the brackets

$D^2 f - D^2 p = d^2 f + d^2 p$ add $D^2 p$ to both sides

$D^2 f = d^2 f + d^2 p + D^2 p$ subtract $d^2 f$ from both sides

$D^2 f - d^2 f = d^2 p + D^2 p$ collect into brackets

$f(D^2 - d^2) = p(d^2 + D^2)$ divide both sides by $D^2 - d^2$

$f = \dfrac{p(d^2 + D^2)}{D^2 - d^2}$

It is very important to understand transposition of formulae since it is very useful for the checking of calculations. Calculations must always be checked. The check can be done either by
(i) doing the calculation in a different order

Example 1.7. Given that $p = mg + \dfrac{mv^2}{r}$ calculate the value of p when $m = 12.84$ kg, $g = 9.81$ ms^{-2}, $v = 16.95$ ms^{-1} and $r = 2.85$ m

CM 12.84 × 9.81 = M+ 16.95 × = × 12.84

÷ 2.85 + RM = 1420.3

Check

CM 16.95 × = ÷ 2.85 × 12.84 = M+ 9.81 × 12.84

+ RM = 1420.3

(ii) reversing the calculation or evaluating one of the other quantities.

Example 1.8. Calculate the value of K in the formula

$K = \dfrac{3n - 2}{n + 1}$ (when $n = 1.84$)

CM 1.84 + 1 = M+ 3 × 1.84 − 2 ÷ RM = 1.239

Check by transposing the formula, we know that

$n + 1 = \dfrac{3n - 2}{K}$

Hence as a check we can calculate $n + 1$, which we know is equal to 2.84.

$$3 \times 1.84 - 2 \div 1.239 = 2.841$$

The reason why we now have 2.841 instead of exactly 2.84 is that the answer 1.239 has been rounded off. To obtain 2.84 exactly in the check we should have retained all of the digits in the answer. That is $K = 1.239\,436\,6$.

Significant errors can be introduced into a calculation if numbers are rounded off or truncated too early in a calculation. When using a calculator the full number of digits should be retained until the end of the calculation. Also one should avoid copying numbers from a calculator in the middle of a calculation, this can also lead to errors.

Example 1.9. Calculate the value of x from the formula $x = \dfrac{dh}{D - d}$ when $d = 1.697$, $D = 1.886$ and $h = 2.917$

(i) if the numbers are rounded off to 2 digits at each stage of the calculation.
(ii) if the numbers are truncated to 2 digits at each stage of the calculation.
(iii) if the full number of digits are retained throughout the calculation.

(i) CM 1.886 − 1.697 = 0.19 (rounded to 2 digits)
     M+ 1.697 × 2.917 = 5.0 (rounded to 2 digits)
     ÷ RM = 26(26.316)
(ii) CM 1.886 − 1.697 = 0.18 (truncated to 2 digits)
     M+ 1.697 × 2.917 = 4.9 (truncated to 2 digits)
     ÷ RM = 27(27.222)
(iii) CM 1.886 − 1.697 = M+ 1.697 × 2.917 =
     ÷ RM = 26.191

In the above example the answers are different if the numbers are rounded off or truncated. It must be remembered that the final answer can only be as accurate as the original data. Hence answers must not be given with strings of digits. The degree of accuracy of an answer is important. All of the digits given in an answer should be meaningful. Practical experience of problems will help with this difficulty. Since calculations can be carried out quickly with a calculator the transposition of a formula can be checked by means of a calculation, consider *Examples 1.4* and

1.5. In *Example 1.4* $A = \pi r^2 \Rightarrow r = \sqrt{\dfrac{A}{\pi}}$

take $r = 1.7$

to find $A : 1.7 \times = \times \pi = 9.079$

to find $r: 9.079 \div \pi = \sqrt{} = 1.7$

In *Example 1.5* $V = \dfrac{2R}{R - r} \Rightarrow R = \dfrac{Vr}{V - 2}$

take $R = 2.9, r = 1.3$

to find $V:$ CM $2.9 - 1.3 = $ M$+ 2 \times 2.9 \div$ RM $= 3.625$

to find $R :$ CM $3.625 - 2 = $ M$+ 3.625 \times 1.3 \div$ RM $= 2.9$

EXERCISE 1.2

Transpose the following formulae in questions 1 to 6

1. $V = \frac{1}{3}\pi R^2 h$ to make R the subject

2. $W = \dfrac{mv^2}{2g}$ to make v the subject

3. $x = \dfrac{dH}{D - d}$ to make d the subject

4. $C = 2\sqrt{2rh - h^2}$ to make r the subject

5. $x = \sqrt{\dfrac{y}{z + y}}$ to make y the subject

6. $f = \dfrac{1}{2\pi\sqrt{LC}}$ to make L the subject

In questions 7 to 9 do the calculation and check your answer by using one of the above mentioned methods.

7. Calculate W from the formula $W = \dfrac{mv^2}{2g}$ when $m = 17.56\,\text{kg}$, $v = 6.95\,\text{ms}^{-1}$ and $g = 9.81\,\text{ms}^{-2}$.

8. If $x = \dfrac{1 - r^2}{1 + r^2}$ calculate the value of x when $r = 0.243$.

9. When $r = 1.87\,\text{m}$ and $h = 2.09\,\text{m}$ calculate the value of C if $C = 2\sqrt{2rh - h^2}$.

1.3 GENERAL EXAMPLES

EXERCISE 1.3

1. Given $f = \dfrac{1}{2\pi\sqrt{LC}}$

(a) Calculate the value of f when $L = 0.000\,12$ and $C = 0.000\,000\,42$

(b) Transpose the formula to make C the subject
(c) Use the transposed formula in (b) to check the answer in (a).

2. Given $E = \dfrac{9KN}{N + 3K}$

(a) Transpose the formula to make K the subject.
(b) Calculate the value of K when $E = 6180$ and $N = 2375$.
(c) Do the calculation a different way to check the answer.

3. Transpose the formula $x = \dfrac{Kgt^2}{2a(1 + K)}$ to make K the subject.

4. Given $I = E\sqrt{(R^2 + w^2 L^2)}$

(a) Calculate the value of I when $E = 2.56, R = 23, w = 2170$ and $L = 0.0196$
(b) Transpose the formula to make R the subject.
(c) Use the transposed formula to check the answer to the calculation of I.

5. Transpose the formula $T = \sqrt{\dfrac{R - S}{S}}$ to make S the subject. Calculate the value of S when $T = 1.74$ and $R = 157$. Check your answer by doing the calculation another way.

6. Transpose the formula $T = 2\pi\sqrt{\dfrac{K^2 + h^2}{gh}}$ to make K the subject.

7. Given $w = \sqrt{\dfrac{1}{LC} - \dfrac{R^2}{4L^2}}$

(a) Calculate the value of w when $L = 0.068, C = 0.000\,087$ and $R = 23.8$.
(b) Transpose the formula to make C the subject.
(c) Use the transposed formula to check the answer to the calculation of w.

8. Given that $B = \dfrac{M + \sqrt{M^2 + T}}{2}$

(a) Transpose the formula to make M the subject.
(b) Calculate the value of M when $B = 12.87$ and $T = 9.76$
(c) Do the calculation in a different way to check the answer.

SELF ASSESSMENT PAPER No 1

Instructions: Answer all the questions

Time allowed: Section A 20 minutes (20 marks)
            Section B 20 minutes (20 marks)

Marks gained: 20+ pass with credit, 16−20 pass,
            less than 16 fail, repeat chapter 1.

Section A

1. Calculate the value of E in the formula $E = \frac{1}{2}mv^2$ when $m = 3.62\,\text{kg}$ and $v = 2.73\,\text{m/s}$.   (2 marks)

2. If $b = \sqrt{\dfrac{a}{a + c}}$ calculate the value of b when $a = 3.67$ and $c = 1.94$.   (4 marks)

3. Calculate the value of p in the formula $p = mg + \dfrac{mv^2}{r}$

for a particle of mass $m = 0.47\,\text{kg}$ and with various values of v and r. ($g = 9.81\,\text{m/s}^2$)

(a) Write out the sequence of steps you would use on your calculator to calculate p. (2 marks)
(b) Write out a different order of steps that could be used to check the calculation. (3 marks)
(c) Complete the table. (3 marks)

r (m)	0.27	0.58	0.73	0.85
v (m/s)	1.86	1.92	1.98	2.03
p				

4. Transpose the formula $V = \frac{4}{3}\pi R^3$ to make R the subject. (2 marks)

5. Transpose the formula $x = \sqrt{\dfrac{a+b}{a-b}}$ to make a the subject. (4 marks)

Section B

1. If $I = \dfrac{bd^3}{12}$, $b = 0.756\,\text{m}$, $d = 0.143\,\text{m}$:

(a) write out the sequence of steps you would use to calculate I and then calculate I.
(b) transpose the formula to make b the subject.
(c) using the transposed formula write out a sequence of steps to check the calculated value of I. Use this sequence of steps to check the value of I obtained in (a).

2. The resonant frequency f_r for a parallel circuit is given by the formula $f_r = \dfrac{1}{2\pi}\sqrt{\dfrac{1}{LC} - \dfrac{R^2}{L^2}}$

(a) Transpose the formula to show that
$$C = \frac{L}{4\pi^2 f_r^2 L^2 + R^2}.$$

(b) Calculate the value of C if $f_r = 53.7$, $L = 0.062$ and $R = 23.8$.

ANSWERS

Exercise 1.1   1. 24 270   2. 0.531,   3. 0.832,
4. 1.34,   5. -0.784,   6. 7.504, 9.664, 12.473,
15.872, 23.071, 26.115,   7. 1.431, 1.816, 2.124,
2.662, 3.333, 5.070,   8. 2.522, 2.798, 2.810,
2.838, 2.861, 2.883,   9. 69.00, 91.43, 117.77,
150.78, 201.73, 249.27,   10. 228.7, 195.6, 172.8,
160.4, 141.4

Exercise 1.2   1. $\sqrt{\dfrac{3V}{\pi h}}$ 2. $\sqrt{\dfrac{2gw}{m}}$

3. $\dfrac{Dx}{x+H}$ 4. $\dfrac{c^2 + 4h^2}{8h}$ 5. $\dfrac{x^2 z}{1-x^2}$

6. $\dfrac{1}{4\pi^2 f^2 C}$ 7. 43.23,   8. 0.8885,

9. 3.714

Exercise 1.3   1. (a) 22 420   (b) $\dfrac{1}{4\pi^2 f^2 L}$

2. (a) $\dfrac{NE}{3(3N-E)}$ (b) 5177

3. $\dfrac{2ax}{gt^2 - 2ax}$ 4. (a) 123.8,

(b) $\sqrt{\dfrac{I^2 - E^2 w^2 L}{E^2}}$ 5. $\dfrac{R}{T^2 + 1}$,   38.98,

6. $\sqrt{\dfrac{T^2 gh - 4\pi^2 h^2}{4\pi^2}}$ 7. (a) 372.6

(b) $\dfrac{4L}{4w^2 L^2 + R^2}$ 8. (a) $\dfrac{4B^2 - T}{4B}$ (b) 12.68

SELF ASSESSMENT PAPER No 1

Section A

1. 13.49,   2. 0.8088,
3. (a) CM $m \times g$ M+ $v \times = \times m \div r$ + RM =,
  (b) CM $v \times = \div r \times m =$ M+ $g \times m = +$ RM =,

  (c) 10.633, 7.598, 7.135, 6.889,   4. $\sqrt[3]{\dfrac{3V}{4\pi}}$

5. $\dfrac{b(x^2 + 1)}{x^2 - 1}$

Section B

		Marks
1. (a) $d \times = \times d \div 12 \times b =$, 0.000 184 m^4		3
(b) $\dfrac{12I}{d^3}$		4
(c) CM $d \times = \times d =$ M+ $12 \times I \div$ RM =, 0.755		3

2. (a) $4\pi^2 f_r^2 = \dfrac{1}{LC} - \dfrac{R^2}{L^2}$

$4\pi^2 f_r^2 L^2 C = L - R^2 C$

$C(4\pi^2 f_r^2 L^2 + R^2) = L$

$C = \dfrac{L}{4\pi^2 f_r^2 L^2 + R^2}$ 6

(b) $C = 0.000\,061\,7$ 4

2 Linear Laws from Experimental Data

2.1 DETERMINATION OF EQUATION FROM THE GRADIENT AND INTERCEPT

When drawing graphs the following points should be remembered:

1. Usually one of the quantities depends on the values of the other quantity. The independent variable is usually plotted on the horizontal axis.

2. The scales for the variables should be carefully chosen so that
(a) the graph fits completely on the paper,
(b) the graph can be easily and accurately read,
use one large square to represent 1, 2, 5, 10 . . . units, avoid using 3, 7, 9, . . . units.

3. The axes should be clearly labelled and if more than one graph is drawn using the same axes then each must be clearly labelled.

4. The points plotted must be clearly marked $\times$, $\odot$, etc. This is important because the points are normally the given information or the information obtained from an experiment. The line drawn is your interpretation of the locus of the points.

In this chapter we are only concerned with straight line graphs. The information will often be obtained from experiments and will seldom be exact. The points may not lie exactly in a straight line but in a band. The best straight line must then be drawn through the points. In level 1 we saw that the equation $y = mx + c$ represents a straight line. When y increases as x increases then m is positive, but when y decreases as x increases then m is negative.

Example 2.1 The speed, v(m/s), of a train after it leaves a station is recorded at time t(s)

v (m/s)	9.0	12.0	14.0	16.5	20.0	22.5
t(s)	2.0	4.0	6.0	8.0	10.0	12.0

(a) Plot a graph of v against t and draw the best straight line through the points.

(b) If the law connecting v and t is $v = mt + c$ determine values for the constants m and c.

(c) Use the law to find (i) the speed when the time is 9 s (ii) the time when the speed is 31 m/s.

(a) see fig. 2.1

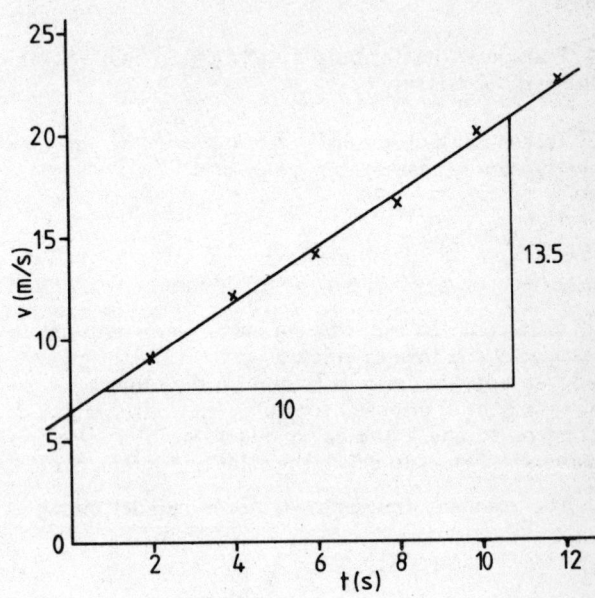

Figure 2.1

(b) From the graph in fig. 2.1 the intercept is 6.3 m/s. Hence $c = 6.3$

The gradient is $\dfrac{13.5}{10} = 1.35$

Hence $m = 1.35$
The law is $v = 1.35t + 6.3$

(c) (i) When $t = 9$ s, $\qquad v = 1.35 \times 9 + 6.3$
$$v = 18.45 \text{ m/s}$$

(ii) When $v = 31$ m/s, $\qquad 31 = 1.35t + 6.3$

$$t = \frac{24.7}{1.35} = 18.3 \text{ s}$$

Example 2.2 A load is applied to a coiled spring so that as the load W(N) increases the length of the spring L (mm) decreases.

W (N)	200	500	650	800	950	1100	1250	1400
L (mm)	78	68	63	58	53	48	43	38

(a) Plot a graph of L against W and draw the best straight line through the points.

(b) If the law connecting L and W is $L = mW + c$ determine the constants m and c.

(c) Use the law to find (i) the length of the spring when the load is 1600 N, (ii) the load when the length of the spring is 45 mm.

(a) See fig. 2.2

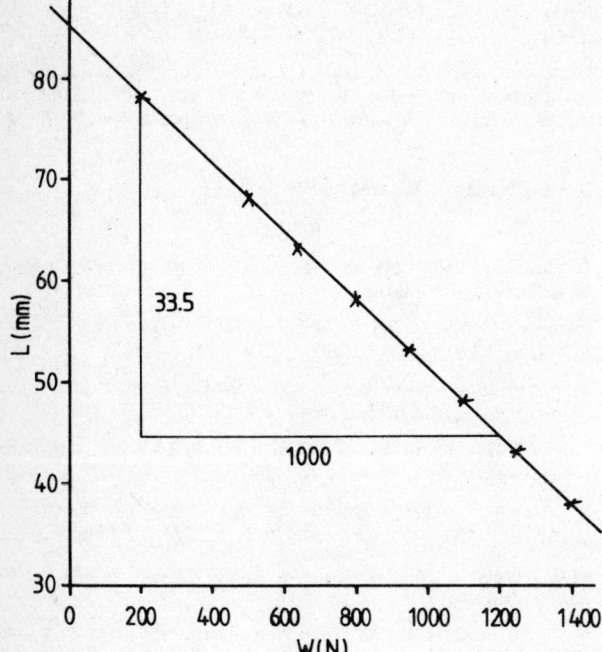

Figure 2.2

(b) From the graph in fig. 2.2 the intercept is 85 mm. Hence $c = 85$.

The gradient is $-\dfrac{33.5}{1000} = -0.0335$

Hence $m = -0.0335$
The law is $L = -0.0335W + 85$
The gradient is negative since the length of the spring decreases as the load increases.

(c) (i) When $W = 1600$ N, $L = -0.0335 \times 1600 + 85$
$$L = 31.4 \text{ mm}$$

(ii) When $L = 45$ mm, $45 = -0.0335W + 85$
$$W = \frac{40}{0.0335} = 1194 \text{ N}$$

EXERCISE 2.1

1. The following table gives the effort E(N) to raise a load W(N) using a lifting machine.

W(N)	100	200	300	450	600	750
E(N)	28	50	74	108	143	178

(a) Plot a graph of E against W and draw the best straight line through the points.
(b) If the law connecting E and W is $E = mW + c$ determine the constants m and c.
(c) Use the law to find (i) the effort required to raise a load of 840 N, (ii) the load raised when an effort of 86 N is applied.

2. The cost of manufacturing components consists of a fixed sum of money £c for setting up the production and a cost per component £m. The total cost £P for producing N components is given in the table below.

N	100	200	400	600	800	1000
P(£)	370	520	820	1120	1420	1720

(a) Plot a graph of P against N and draw the best straight line through the points.
(b) If the law connecting P and N is $P = mN + c$ determine the constants m and c.
(c) Use the law to find (i) the cost of manufacturing 480 components, (ii) the number of components produced if the total cost is £2000.

3. In a controlled cooling process, the temperature $T(^\circ \text{C})$ is recorded at times t(s)

t(s)	20	40	60	80	100	120
$T(^\circ \text{C})$	230	198	162	130	92	60

(a) Plot a graph of T against t and draw the best straight line through the points.
(b) If the law connecting T and t is $T = mt + c$ determine the constants m and c.
(c) Use the law to find (i) the temperature after 135 s, (ii) the time when the temperature was 178°C.

4. The following table gives the results of a test on two metallic filament lamps connected in parallel.

V (volts)	60	65	70	75
I (amperes)	0.55	0.57	0.59	0.61

(a) Calculate the values of the resistance R, by using $R = \dfrac{V}{I} \ \Omega$.
(b) Plot a graph of R against V and draw the straight line through the points.
(c) If the law connecting R and V is $R = aV + b$ determine the constants a and b.

2.2 DETERMINATION OF EQUATION FROM TWO POINTS ON THE LINE

The constants in the equation can be found by taking two points from the line and solving a pair of simultaneous equations. Sometimes this method must be used since the intercept cannot always be read from the graph. When this method is used the two points must be taken from the line of the graph, they must not be taken from the table of values since these points may not lie exactly on the line.

Example 2.3 In a friction test, the force F(N) required to move a load W(N) was measured for various loads.

W(N)	30	35	40	50	60	70
F(N)	9.7	10.4	11.3	12.7	13.8	15.4

Draw the graph of F against W and from it determine the constants m and c for the law $F = mW + c$.

The two points $P(W = 40, F = 11.2)$ and $Q(W = 60, F = 13.9)$ are taken from the graph.
The law is $F = mW + c$
Hence when $W = 40$ and $F = 11.2$ then $11.2 = 40m + c$
$$(1)$$
and when $W = 60$ and $F = 13.9$ then $13.9 = 60m + c$
$$(2)$$

$40m + c = 11.2$ (1)
$60m + c = 13.9$ (2)
Subtract (1) from (2) $20m = 2.7$
$$\text{or } m = \frac{2.7}{20} = 0.135$$

Substitute $m = 0.135$ into (1) $40 \times 0.135 + c = 11.2$
$$\text{or } 5.4 + c = 11.2$$
$$c = 11.2 - 5.4 = 5.8$$

The law is $F = 0.135W + 5.8$

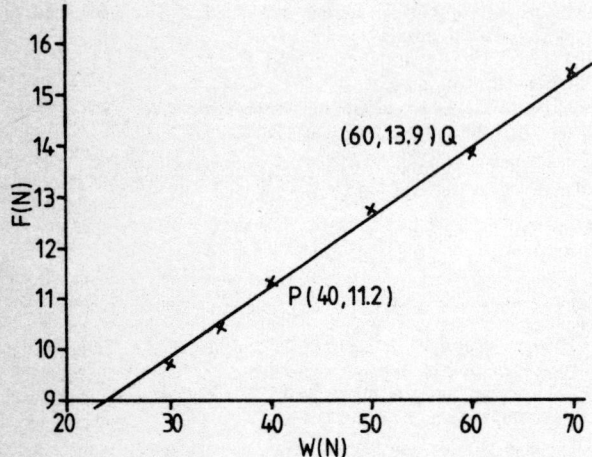

Figure 2.3

EXERCISE 2.2

1. The following table shows the results from a test on an electrical machine.

Current, I(A)	7.4	8.0	9.8	11.0	11.8	12.8
Resistance, R(Ω)	3.6	4.3	4.7	5.7	6.6	7.2

Draw the graph of I against R and from it determine the constants a and b for the law $I = aR + b$.

2. Tests were carried out on a metal to find the relationship between the Brinell hardness number H and the tensile strength T(MN/m²)

Hardness number, H	112	119	135	147	168
Tensile strength, T(MN/m²)	408	431	477	514	585

Draw the graph of T against H and from it determine the constants m and c for the law $T = mH + c$.

3. The resistance of a conductor changes with temperature according to the relationship $R = R_0(1 + \alpha t)$ where R is the resistance of the conductor at $t°$ C, R_0 is the resistance of the conductor at 0°C and α is the temperature coefficient of resistance. During an experiment to determine the temperature coefficient of resistance of a copper coil the following observations were made:

R(Ω)	5.45	5.50	5.60	5.65	5.75
$t(°$ C)	19.5	22.0	27.5	30.0	35.5

Draw the graph of R against t and from it determine the constants a and b for the law $R = b + at$. Hence find the values of R_0 and α when this law is written $R = R_0 + R_0\alpha t$.

2.3 GENERAL EXAMPLES

EXERCISE 2.3

1. The table below shows the results obtained when a power test was made on a lathe.

n (rev/s)	1.8	3.1	4.9	7.7	10.6	14.2
P (Watts)	1370	1790	2360	3250	4160	5300

Plot a graph of these values and from the graph find the law connecting P and n in the form $P = an + b$.

2. The following values of x and y satisfy an equation of the form $y = ax + b$.

x	0.063	0.071	0.079	0.089	0.098	0.108
y	95	90	85	78	72	66

Plot a graph of y against x and determine the values of the constants a and b.

3. In a friction test the force F(N) required to move a load W(N) was measured and recorded in the table below.

W(N)	30	33	35	37	40	43	45
F(N)	8.7	9.4	9.9	10.4	11.2	11.9	12.4

Plot a graph of F against W and from it determine the values of the constants a and b for the law $F = aW + b$.

SUMMARY

1. The equation of a straight line can be written $y = mx + c$ where m is the gradient and c is the intercept on the y-axis.
2. When y increases as x increases the gradient is positive. When y decreases as x increases the gradient is negative.
3. When recorded or experimental points are plotted they may not lie exactly on a straight line. The best straight line is drawn through the points with some points above the line and some below the line.
4. The values of m and c in the equation $y = mx + c$ can be found by taking two points from the line and solving a pair of simultanaeous equations.

SELF ASSESSEMENT PAPER No 2

Instructions: Answer both questions
Time allowed: 30 minutes (20 marks each question)
Marks Gained: 20+ pass with credit, 16–20 pass, less than 16 fail, repeat chapter 2.

1. A test carried out on a screw and nut mechanism produced the following results, the axial load on the screw being W and the corresponding effort E.

W(N)	0	800	1600	2400	3200	4000	4800
E(N)	8.9	26.3	37.8	51.6	63.2	78.3	90.8

Plot these values and find the constants a and b if the law of the mechanism is $E = aW + b$.

2. The following table of values shows how the length of a wire x(m) varies when subjected to a load W(g).

x(m)	0.1306	0.1345	0.1370	0.1424	0.1462	0.1500
W(g)	140	160	180	200	220	240

Draw a graph to show that these satisfy a law of the form $W = ax + b$. From the graph determine values of a and b, by selecting two points from the graph and solving a pair of simultaneous equations.

ANSWERS

Exercise 2.1 1. (b) $m = 0.23$, $c = 5$, (c) (i) 198 N,
(ii) 352 N, 2. (b) $m = 1.5$, $c = 220$, (c) (i) £940,
(ii) 1186 3. (b) $m = -1.7$, $c = 264$,
(c) (i) 35° C, (ii) 50.6 s 4. $a = 0.95$, $b = 52.3$.

Exercise 2.2 1. a 1.5, $b = 2.57$, 2. $m = 3.13$,
$c = 57$, 3. $a = 0.0185$, $b = 5.1$, $R_0 = 5.1$,
$\alpha = 0.0036$.

Exercise 2.3 1. $P = 315n + 800$, 2. $a = -644$,
$b = 136$, 3. $a = 0.24$, $b = 1.52$.

SELF ASSESSMENT PAPER No 2

	Marks
1. Graph	9
From graph when $W = 0$, $E = 10.5$	
hence $b = 10.5$	4
Gradient of graph $= \dfrac{50}{3000} = 0.017$	5
$a = 0.017$	1
The law is $E = 0.017\,W + 10.5$	1
2. Graph	8
Two points from the graph:	
$x = 0.135$, $W = 162$ g; $x = 0.15$, $W = 240$ g	2
$162 = 0.135a + b$ (1)	1
$240 = 0.150a + b$ (2)	1
Subtract (1) from (2) $78 = 0.015a$	2
$a = 5200$	1
Substitute this into (1)	
$162 = 0.135 \times 5200 + b$	2
$b = -540$	2
Law is $W = 5200x - 540$	1

3 Graphs of Algebraic Equations

3.1 SIMULTANEOUS LINEAR EQUATIONS

When a problem involves two unknown quantities it is necessary to have two equations to solve it. The simplest case is when both equations represent straight lines. The solution is found by drawing the straight line graphs, representing the two lines, and reading off the values of x and y at the point of intersection.

Example 3.1 Solve graphically, the simultaneous equations $y = 2x - 2$ and $y = -3x + 9$

When drawing a straight line graph it is only necessary to calculate two points but in practice it is better to find 3 points, the third point acting as a check on the working.

$y = 2x - 2$ $x =$ 0   2   4
                      $y = -2$ 2   6
$y = -3x + 9$ $x =$ 0   2   4
                      $y =$ 9   3   −3

From these values of x and y the two lines are drawn on the graph in fig. 3.1.

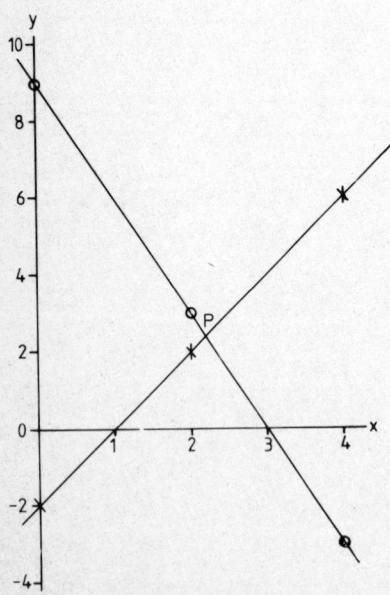

Figure 3.1

These two lines meet at the point P (2.2, 2.4). Hence the solution to these simultaneous equations is $x = 2.2$, $y = 2.4$.

Example 3.2 Solve graphically the simultaneous equations $5y + 2x = 24$, and $8y - 4x = 9$.
To find the values of x and y to draw the graphs we will first rewrite these equations.

(1) $5y + 2x = 24$, $5y = 24 - 2x$, $y = \dfrac{24 - 2x}{5}$

$x = 0$ 5   10
$y = 4.8$ 2.8   0.8

(2) $8y - 4x = 9$, $8y = 9 + 4x$, $y = \dfrac{9 + 4x}{8}$

$x = 0$ 5   10
$y = 1.1$ 3.6   6.1

From these values of x and y the graphs of the straight lines are drawn in fig. 3.2.

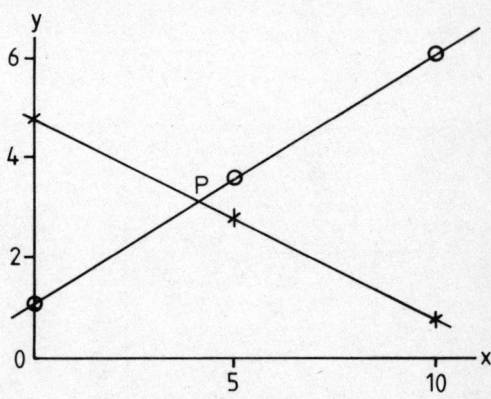

Figure 3.2

These two lines meet at the point P (4.0, 3.2). Hence the solution to these simultaneous equations is $x = 4.0$, $y = 3.2$.

EXERCISE 3.1

Solve graphically the following simultaneous equations.

1. $y = 3x - 5$ 2. $y = 3x + 2$ 3. $y + x = 4$
    $y = 3 - x$ $y = -2x + 10$ $y - 2x = 7$
4. $2x + 2y = 12$
    $x + 3y = 11$

3.2 QUADRATIC EQUATIONS

The equations in 3.1 were all linear equations, that is they represented straight lines. They were also of the first degree, that is the powers of x and y were always one. If one or both of the powers are two the equations are of the second degree and are called quadratic equations. We will be considering here equations of the form $y = ax^2 + bx + c$, where a, b and c are constants. The graphs of these quadratic equations are curves, they all have one point where they turn and are called parabolas.

Example 3.3 Plot the graph of $y = 2x^2 - 3x - 8$ between $x = -2$ and $x = 4$. From the graph write down the values of x which satisfy the equation $2x^2 - 3x - 8 = 0$

x	-2	-1	0	1	2	3	4
$2x^2$	8	2	0	2	8	18	32
$-3x$	6	3	0	-3	-6	-9	-12
-8	-8	-8	-8	-8	-8	-8	-8
y	6	-3	-8	-9	-6	1	12

The graph is drawn in fig. 3.3

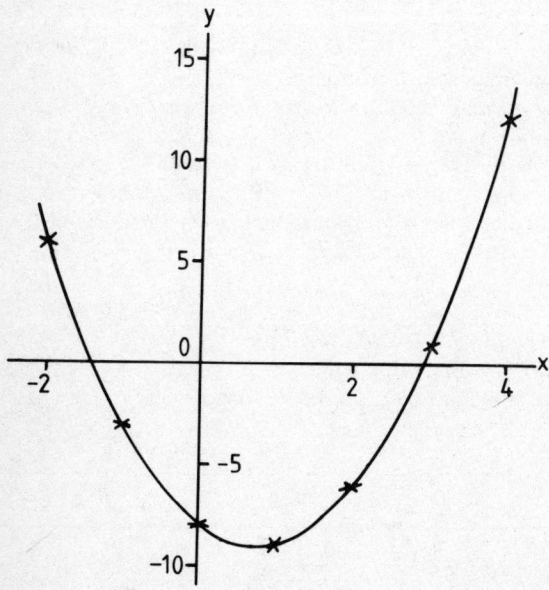

Figure 3.3

The question also asks what values of x will make $2x^2 - 3x - 8 = 0$. The graph is $y = 2x^2 - 3 - 8$ and so when $y = 0$ then $2x^2 - 3x - 8 = 0$. $y = 0$ along the x-axis and hence the values of x which make $2x^2 - 3x - 8 = 0$ will be where the curve meets the

x-axis. Hence the values of x which satisfy $2x^2 - 3x - 8 = 0$ are $x = -1.4$ and $x = 2.9$.

Example 3.4 Plot the graph of $y = 2x^2 + 7x + 3$ between $x = -5$ and $x = 2$. Find the values of x which satisfy the equation $2x^2 + 7x - 7 = 0$.

x	-5	-4	-3	-2	-1	0	1	2
$2x^2$	50	32	18	8	2	0	2	8
$7x$	-35	-28	-21	-14	-7	0	7	14
3	3	3	3	3	3	3	3	3
y	18	7	0	-3	-2	3	12	25

The graph is drawn in fig. 3.4

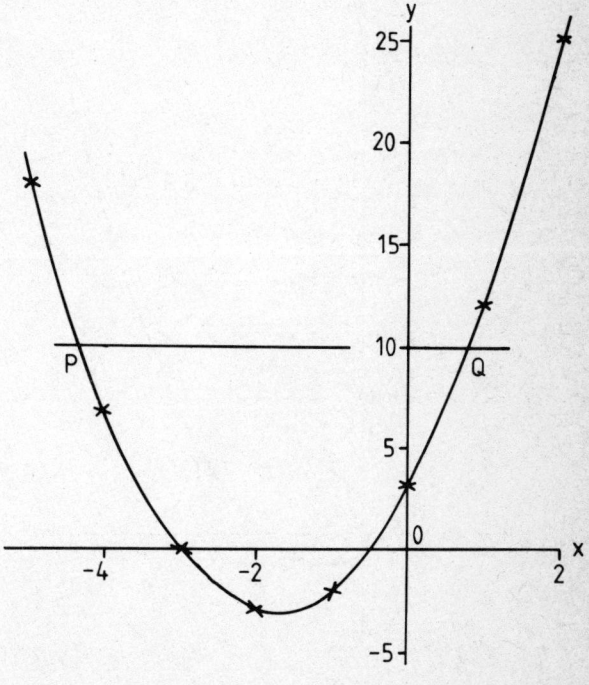

Figure 3.4

The question asks for the values of x which satisfy the equation $2x^2 + 7x - 7 = 0$. This can be written $2x^2 + 7x - 7 + 10 = 10$ or $2x^2 + 7x + 3 = 10$
The values of x which satisfy this equation are given by the points where the line $y = 10$ meets the curve $y = 2x^2 + 7x + 3$. These are the points P and Q in fig. 3.2. The values of x which satisfy the equation $2x^2 + 7x - 7 = 0$ are $x = -4.3$ and $x = 0.8$.

Example 3.5 Draw the graphs of $y = 2x^2 - 2x - 1$ and $y = 3x + 2$ between $x = -2$ and $x = 4$. From the graph find the values of x and y which satisfy the equations $y = 2x^2 - 2x - 1$ and $y = 3x + 2$.

x	-2	-1	0	1	2	3	4
$2x^2$	8	2	0	2	8	18	32
$-2x$	4	2	0	-2	-4	-6	-8
-1	-1	-1	-1	-1	-1	-1	-1
y	11	3	-1	-1	3	11	23

$y = 3x + 2$
$x = -2 \quad y = -4$
$x = 0 \quad\;\; y = 2$
$x = 4 \quad\;\; y = 14$

The graphs are drawn in fig. 3.5

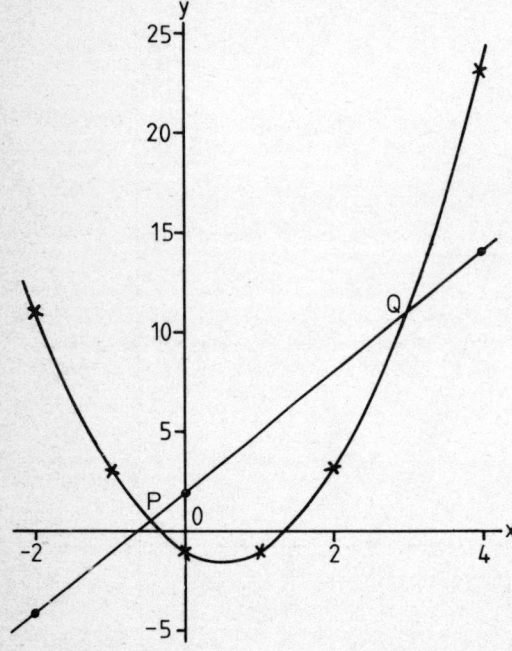

Figure 3.5

The values of x and y which satisfy the equations $y = 2x^2 - 2x - 1$ and $y = 3x + 2$ are given by the points P and Q where the line meets the curve.
The solutions are $x = -0.5, \quad y = 0.5$
$\qquad\qquad\qquad\quad x = 3.0, \quad\;\; y = 11.0$

Where the line meets the curve the values of y are the same and hence $2x^2 - 2x - 1 = 3x + 2$ or $2x^2 - 5x - 3 = 0$.
Therefore the two values of x are the solutions of this equation. The solutions of $2x^2 - 5x - 3 = 0$ are $x = -0.5$ and $x = 3.0$.

EXERCISE 3.2

1. Draw the graph of $y = 2x^2 - 5x - 10$ between $x = -2$ and $x = 4$. From the graph write down the values of x which satisfy the equation $2x^2 - 5x - 10 = 0$.

2. Draw the graph of $y = -2x^2 + 4x + 3$ between $x = -2$ and $x = 4$. From the graph write down the values of x which satisfy the equation $-2x^2 + 4x + 3 = 0$.

3. Draw the graph of the equation $y = x^2 - 2x - 3$ for values of x from -2 to $+5$. From the graph find the solutions to the equations $x^2 - 2x - 3 = 0$ and $x^2 - 2x - 1 = 0$.

4. Draw the graph of $y = x^2 - 3x - 2$ for values of x from -2 to $+5$. From the graph find the solutions to the equations $x^2 - 3x - 2 = 0$ and $x^2 - 3x - 4 = 0$.

5. Draw the graph of $y = x^2 - 4$ between $x = -4$ and $x = 4$. Plot on the same graph $y = 2x + 1$ and hence solve the simultaneous equations $y = x^2 - 4$ and $y = 2x + 1$. Write down the quadratic equations in x which can also be solved by these graphs.

6. Draw the graph of $y = x^2 - 4x + 7$ between $x = 0$ and $x = 5$. Plot on the same graph the line $y = x + 1$ and hence solve the simultaneous equations $y = x^2 - 4x + 7$ and $y = x + 1$. Write down the quadratic equation in x which can also be solved from these graphs.

3.3 THE CUBIC EQUATION

When the equation contains x^3 it is of the third degree and it is called a cubic equation. The graph of the equation $y = ax^3 + bx^2 + cx + d$, where a, b, c and d are constants has two points where the curve turns. The curve may cut the x-axis three times and then the equation $ax^3 + bx^2 + cx + d = 0$ has three solutions. If both turning points are above the x-axis or both turning points are below the x-axis then the curve will only cut the x-axis once and hence the equation will only have one solution.

Example 3.6 Draw the graph of $y = x^3 - 3x^2 - x + 3$ between the values of x from -2 to $+4$. Write down the solutions to $x^3 - 3x^2 - x + 3 = 0$.

x	-2	-1	0	1	2	3	4
x^3	-8	-1	0	1	8	27	64
$-3x^2$	-12	-3	0	-3	-12	-27	-48
$-x$	2	1	0	-1	-2	-3	-4
$+3$	3	3	3	3	3	3	3
y	-15	0	3	0	-3	0	15

The graph is drawn in fig. 3.6

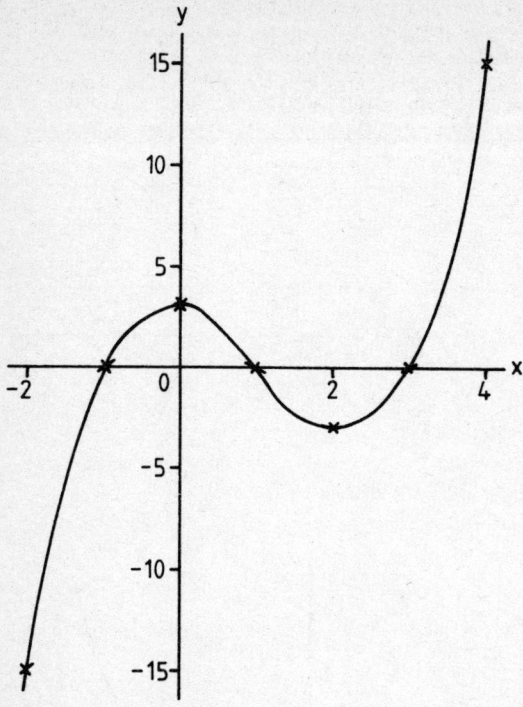

Figure 3.6

The solutions to $x^3 - 3x^2 - x + 3 = 0$ are given by the points where the curve cuts the x-axis. The solutions are $x = -1$, $x = 1$ and $x = 3$.

Example 3.7 Draw the graph of $y = x^3 - x^2 - x - 2$ between the values of x from -1 to $+3$. Write down the solutions to $x^3 - x^2 - x - 2 = 0$.

x	-1	0	1	2	3
x^3	-1	0	1	8	27
$-x^2$	-1	0	-1	-4	-9
$-x$	1	0	-1	-2	-3
-2	-2	-2	-2	-2	-2
y	-3	-2	-3	0	13

The graph is drawn in fig. 3.7

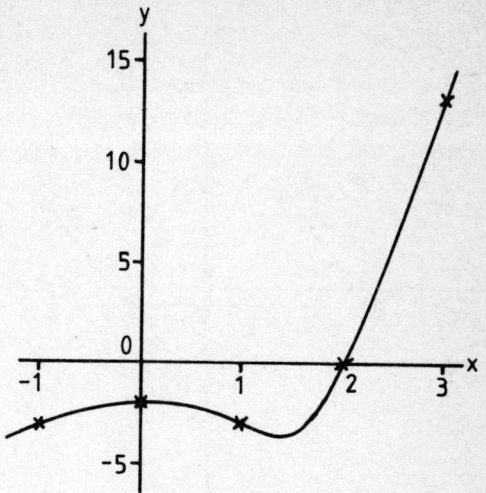

Figure 3.7

The solutions to $x^3 - x^2 - x - 2 = 0$ are given by the points where the curve cuts the x-axis. Because both turning points are below the x-axis there is only one solution to this equation. The solution to this equation is $x = 2$.

Example 3.8 Draw the graph of $y = 3x^3 + 4x^2 - 13x + 3$ for values of x from $x = -4$ to $x = +2$, and hence find the solutions to the equations
(a) $3x^3 + 4x^2 - 13x + 6 = 0$,
(b) $3x^3 + 4x^2 - 13x + 17 = 0$.

x	-4	-3	-2	-1	0	1	2
$3x^3$	-192	-81	-24	-3	0	3	24
$4x^2$	64	36	16	4	0	4	16
$-13x$	52	39	26	13	0	-13	-26
3	3	3	3	3	3	3	3
y	-73	-3	21	17	3	-3	17

The graph is drawn in fig. 3.8

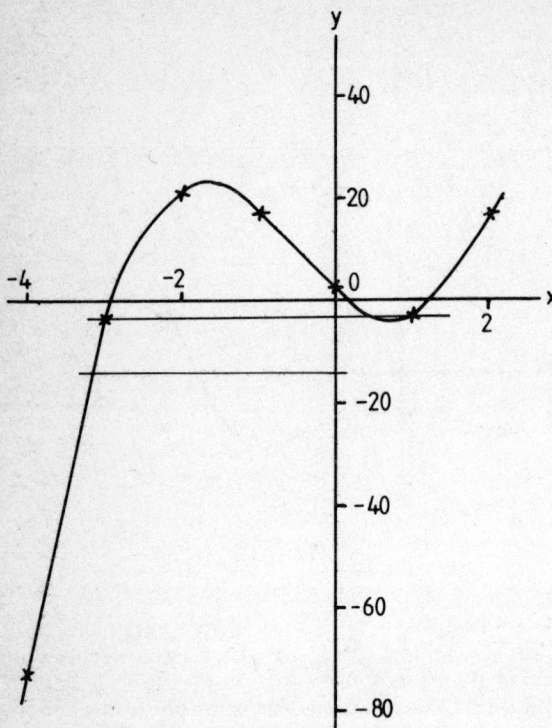

Figure 3.8

(a) The equation $3x^3 + 4x^2 - 13x + 6 = 0$ can be written $3x^3 + 4x^2 - 13x + 3 = -3$. Hence the solutions of the equation are where the curve cuts the line $y = -3$. They are $x = -3$, $x = 0.5$ and $x = 1.0$.
(b) The equation $3x^3 + 4x^2 - 13x + 17 = 0$ can be written $3x^3 + 4x^2 - 13x + 3 = -14$. Hence the solutions of the equation are where the curve cuts the line $y = -14$. Solution is $x = -3.2$.

EXERCISE 3.3

1. Draw the graph of $y = x^3 + 5x^2 - 2x - 24$ between $x = -5$ and $x = 3$. Write down the solutions to $x^3 + 5x^2 - 2x - 24 = 0$.

2. Draw the graph of $y = 2x^3 - 7x - 9$ between $x = -3$ and $x = 4$. Write down the solutions to $2x^3 - 7x - 9 = 0$.

3. Solve graphically the equation $x^3 - 12x - 4 = 0$ by drawing the graph of $y = x^3 - 12x - 4$ between $x = -4$ and $x = 4$.

4. Draw the graph of $y = x^3 - 3x^2 - 6x + 12$ between $x = -3$ and $x = 4$. Hence determine the solutions of (a) $x^3 - 3x^2 - 6x + 12 = 0$ and (b) $x^3 - 3x^2 - 6x^2 + 8 = 0$.

5. On the same axes and with the same scales plot the graphs of $y = 3 + 2x - x^2$ and $y = \dfrac{2}{x}$ from $x = -3$ to $x = 3$.

Show that the points of intersection of these two graphs are the solutions to the equation $x^3 - 2x^2 - 3x + 2 = 0$. Write down the solutions to the equation $x^3 - 2x^2 - 3x + 2 = 0$.

3.4 GENERAL EXAMPLES

Example 3.9 On the same axes and with the same scales plot the graphs of $y = 2x^2 - x - 1$ and $y = x + 2$. From your graphs solve the equations (a) $2x^2 - 2x - 3 = 0$, (b) $2x^2 - x = 0$.

x	-2	-1	0	1	2	3
$2x^2$	8	2	0	2	8	18
$-x$	2	1	0	-1	-2	-3
-1	-1	-1	-1	-1	-1	-1
y	9	2	-1	0	5	14

$y = x + 2$
$x = -2$, $\quad y = 0$
$x = 0$, $\quad y = 2$
$x = 3$, $\quad y = 5$

The graphs are shown in fig. 3.9

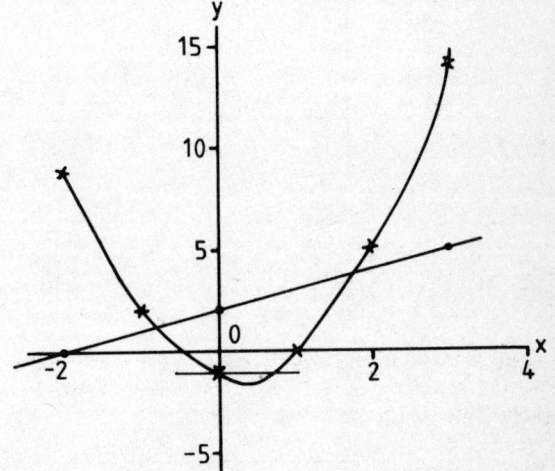

Figure 3.9

(a) The equation $2x^2 - 2x - 3 = 0$ can be written $2x^2 - x - 1 = x + 2$. The solutions to this equation are where the two graphs cut. The solutions to $2x^2 - 2x - 3 = 0$ are $x = -0.85$ and $x = 1.85$.
(b) The equation $2x^2 - x = 0$ can be written $2x^2 - x - 1 = -1$. The solutions to this equation are therefore where the line $y = -1$ cuts the curve $y = 2x^2 - x - 1$. The solutions to the equation $2x^2 - x = 0$ are $x = 0$ and $x = 0.7$.

EXERCISE 3.4

1. Solve graphically the simultaneous equations $2x + y = 7$, $x - 5y = 9$.

2. A projectile rises according to the formula $h = 250t - 5t^2$. Draw a graph for values of t from 0 to 50 s. From the graph find the times when the projectile is at a height of 2800 m.

3. On the same axes and with the same scales plot the graphs of $y = x^2 - x + 1$ and $y = x + 3$. From your graphs solve the equations (a) $x^2 - 2x - 2 = 0$, (b) $x^2 - x - 2 = 0$.

4. Draw the graph of $y = 2x^3 - 3x^2 - 11x + 6$ between the values of $x = -3$ and $x = 4$. From your graph find the solutions to the equations
(a) $2x^3 - 3x^2 - 11x + 6 = 0$
(b) $2x^3 - 3x^2 - 11x - 6 = 0$.
Draw on the graph the line $y = 4x + 4$ and hence find the solutions to the equation $2x^3 - 3x^2 - 15x + 2 = 0$.

SUMMARY

1. A pair of simultaneous linear equations can be solved by plotting their graphs and finding the point of intersection.
2. A quadratic equation can be solved by drawing a graph of the function and finding the values of x where the graph cuts the x-axis.
3. A pair of simultaneous equations, one linear and one quadratic can be solved by drawing their graphs and finding the points of intersection.
4. A cubic equation can be solved by drawing a graph of the function and finding the points where it cuts the x-axis. There may be one or three solutions.

SELF ASSESSMENT PAPER No 3

Instructions: Answer all the questions
Time allowed: One hour (question 1, 14 marks;
questions 2 and 3, 23 marks each)
Marks gained: 30+ marks pass with credit, 24–30 marks pass, less than 24 marks fail, repeat chapter 3.

1. Solve graphically the simultaneous equations $5y + 2x = 12$ and $y - x = 1$.

2. Draw the graph of $y = x^2 - 4x + 3$ for values of x between $x = -1$ and $x = 5$. Write down the solutions to the equation $x^2 - 4x + 3 = 0$. On the same graph draw the straight line $y = 2x - 2$ and hence find the solution to the simultaneous equations $y = x^2 - 4x + 3$ and $y = 2x - 2$.

3. Draw the graph of $y = x^3 - 3x^2 - 6x + 8$ between $x = -4$ and $x = 4$. Hence find the solutions to the equations
(a) $x^3 - 3x^2 - 6x + 8 = 0$, (b) $x^3 - 3x^2 - 6x + 4 = 0$.

ANSWERS

Exercise 3.1 1. $x = 2, y = 1$, 2. $x = 1.6$, $y = 6.8$, 3. $x = -1, y = 5$, 4. $x = 2$, $y = 3$.

Exercise 3.2 1. $3.8, -1.3$, 2. $2.6, -0.6$, 3. $-1, 3; 0.5, 2.4$ 4. $-0.55, 3.6; -1, 4$, 5. $x = -1.4, y = -1.8; x = 3.4, y = 7.8$, $x^2 - 2x - 5 = 0$, 6. $x = 2, y = 3; x = 3, y = 4$, $x^2 - 5x + 6 = 0$.

Exercise 3.3 1. $2, -3, -4$, 2. 3.1, 3. $-3.2, -0.3, 3.7$, 4. (a) $-2.2, 1.4, 3.7$, (b) $-2, 1, 4$, 5. $-1.4, 0.6, 2.8$.

Exercise 3.4 1. $x = 4, y = -1$, 2. 17 s, 33 s, 3. (a) $2.7, -0.7$, (b) $2, -1$, 4. (a) $-2, 0.5, 3$, (b) $-1, -1, 3.6, -2.2, 0.14, 3.7$.

SELF ASSESSMENT PAPER No. 3.

	Marks
1. $y = \dfrac{12 - 2x}{5}$	1
$x = 0, y = 2.4; x = -2, y = 3.2; x = 3,$ $y = 1.2$	3
$y = x + 1$	1
$x = 0, y = 1; x = -2,$ $y = -1; x = 3, y = 4$	3
Graphs	4
$x = 1, y = 2$	2

2.
x	-1	0	1	2	3	4	5	
y	8	3	0	-1	0	3	8	7

	Marks
Graph	5
$x = 1, x = 3$	2
$y = 2x - 2; x = -1, y = -4; x = 0, y = -2;$ $x = 5, y = 8$	3
Drawing line on graph	2
$x = 1, y = 0$ and $x = 5, y = 8$	4

3.
x	-4	-3	-2	-1	0	1	2	3	4	
y	-80	-28	0	10	8	0	-8	-10	0	9

	Marks
Graph	6
(a) $x = -2, 1, 4$	2
(b) Line $y = 4$	3
$x = -1.6, 0.5, 4.4$	3

PHASE TEST 1

Instructions: Answer all questions in section A and section B
Time allowed: Section A 20 minutes (20 marks)
Section B 40 minutes (20 marks each question)
Marks gained: 30+ pass with credit, 24–30 pass, less than 24 fail repeat chapters 1, 2 and 3.

Section A

1. Calculate the volume of a cap of a sphere $V = \dfrac{\pi h^2}{3} (3R - h)$ when the radius $R = 127$ mm and the height $h = 80$ mm.

2. The curved surface area of a frustum of a cone is given by $A = \pi(r + R)l$. Transpose the formula to make r the subject.

3. Write down the gradient of the line $2y = 3 - x$ and also the intercept this line makes on the y-axis.

Section B

1. Given $f = \dfrac{1}{2\pi\sqrt{LC}}$

(a) Calculate the value of f when $L = 1.3 \times 10^{-4}$ and $C = 3.82 \times 10^{-7}$.

(b) Transpose the formula to make C the subject.

(c) Use the transposed formula in (b) to check the answer in (a).

2. A test on a filament lamp gave the following values of voltage for various values of resistance R.

R (ohms) 47   63   86   117   139
V (volts) 20   34   52   77   96

Plot a graph of R against V and determine the constants a and b if the law is $R = aV + b$.

3. Draw the graph of $y = 2x^2 - 4x - 8$ for values of x from -2 to 4. On the same graph and with the same scales draw the graph of $y = 3x - 4$. Hence find the solutions to the simultaneous equations $y = 2x^2 - 4x - 8$ and $y = 3x - 4$.

ANSWERS PHASE TEST 1

Marks

Section A

1. $V = \dfrac{\pi \times 80^2}{3}(3 \times 127 - 80)$ — 2

$V = 201\,700 \text{ mm}^3$ — 4

2. $A = \pi rl + \pi Rl$ — 2

$\pi rl = A - \pi Rl$ — 3

$r = \dfrac{A - \pi Rl}{\pi l}$ — 3

3. $y = -\frac{1}{2}x + \frac{3}{2}$ — 2
gradient $= -\frac{1}{2}$, intercept $= \frac{3}{2}$ — 4

Section B

1. (a) $f = \dfrac{1}{2 \times \pi \times \sqrt{1.3 \times 10^{-4} \times 3.82 \times 10^{-7}}}$ — 2

$f = 2.26 \times 10^4$ — 5

(b) $2\pi f\sqrt{LC} = 1$ — 2

$4\pi^2 f^2 LC = 1$ — 2

$C = \dfrac{1}{4\pi^2 f^2 L}$ — 2

(c) $C = \dfrac{1}{4\pi^2 \times (2.26 \times 10^4)^2 \times 1.3 \times 10^{-4}}$ — 2

$C = 3.81 \times 10^{-7}$ — 5

2. Graph of R vertically against V — 8
Intercept on R axis $= b = 24$ — 6
Gradient $= a = 1.2$ — 6

3. $y = 2x^2 - 4x - 8$

x	-2	-1	0	1	2	3	4
y	8	-2	-8	-10	-8	-2	8

$y = 3x - 4$

x	-2	0	4
y	-10	-4	8

— 4

— 2

Graph of $y = 2x^2 - 4x - 8$ — 6
Graph of $y = 3x - 4$ — 2
Values of x, where the line cuts the curve are the solutions of the equations. $x = 4$ or $x = -\frac{1}{2}$ — 6

4 Radians and Parts of a Circle

4.1 RADIANS

One radian is the angle subtended at the centre of a circle of radius r by an arc of length r. The angle at the centre of a circle is proportional to the arc on which it stands. Hence an arc of length πr will subtend an angle π radians at the centre of the circle. But πr is the length of a semicircular arc which subtends an angle of $180°$ at the centre.

π radians $= 180°$
or 1 radian $\approx 57.3°$

When a wheel is rotating at n revolutions per second this is equivalent to $2\pi n$ radians per second or $360n$ degrees per second.

Example 4.1 Write down the following angles in radians,

(a) $60°$,   (b) $45°$,   (c) $50°36'$ (d) $258°17'$

(a) $60° = 60 \times \dfrac{\pi}{180} = \dfrac{\pi}{3}$ rad $= 1.0472$ rad

(b) $45° = 45 \times \dfrac{\pi}{180} = \dfrac{\pi}{4}$ rad $= 0.7854$ rad

(c) $50°36' = 50.6 \times \dfrac{\pi}{180} = 0.8831$ rad

(d) $258°17' = 258.283 \times \dfrac{\pi}{180} = 4.5079$ rad

For angles such as $30°, 45°, 60°$ etc the radian measure is usually expressed as a fraction of π.
Tables are available for changing degrees to radians but the calculation is usually done on a calculator using the conversion factor $\dfrac{\pi}{180}$.

Example 4.2 Write the following angles in degrees and minutes.

(a) π rad,   (b) $\dfrac{\pi}{2}$ rad,   (c) $\dfrac{\pi}{6}$ rad   (d) 3.85 rad

(a) π rad $= \pi \times \dfrac{180}{\pi} = 180°$

(b) $\dfrac{\pi}{2}$ rad $= \dfrac{\pi}{2} \times \dfrac{180}{\pi} = 90°$

(c) $\dfrac{\pi}{6}$ rad $= \dfrac{\pi}{6} \times \dfrac{180}{\pi} = 30°$

(d) 3.85 rad $= 3.85 \times \dfrac{180}{\pi} = 220.59° = 220°35'$

Example 4.3 A wheel is rotating at 9.63 rad/s. Express this in rev/s and also calculate the angle through which the wheel turns, in degrees per second.

9.63 rad/s $= \dfrac{9.63}{2\pi}$ rev/s $= 1.53$ rev/s

9.63 rad $= 9.63 \times \dfrac{180}{\pi} = 551.76° = 551°45'$

9.63 rad/s is equivalent to 1.53 rev/s
The angle turned through in one second is $551°45'$

EXERCISE 4.1

1. Write down the following angles in radians.
(a) $90°$ (b) $30°$ (c) $135°$ (d) $27°18'$
(e) $137°29'$ (f) $309°47'$

2. Write down the following angles in degrees and minutes.

(a) $\dfrac{\pi}{3}$ rad   (b) $\dfrac{\pi}{4}$ rad   (c) $\dfrac{3\pi}{2}$ rad   (d) 1.75 rad

(e) 2.96 rad   (f) 3.65 rad.

3. A wheel is rotating at 15.74 rad/s. Express this in rev/s and also calculate the angle through which the wheel turns in one second.

4. Calculate the number of radians turned through in 8 s by a shaft rotating at 2 rev/s.

5. A shaft turns through an angle of $736°$ in one second. Express this in rev/s and in rad/s.

6. A wheel is rotating at 620 revolutions per minute. Find the angle turned through in one second (a) in radians and (b) in degrees.

4.2 ARC LENGTH OF A CIRCLE

The length of an arc of a circle is proportional to the angle subtended at the centre of the circle by the arc. The circumference of a circle is $2\pi r$.

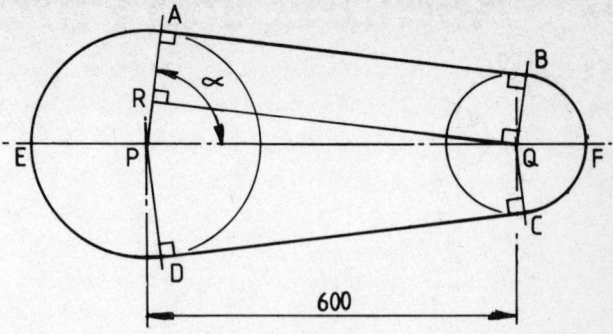

Figure 4.2

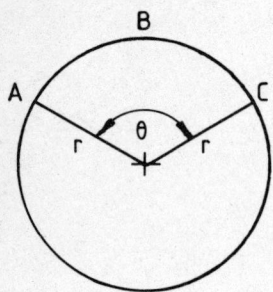

Figure 4.1

$$\frac{\text{Arc length ABC}}{\text{Circumference of circle}} = \frac{\theta^\circ}{360^\circ} = \frac{\theta \text{ rad}}{2\pi \text{ rad}}$$

Working in radians, $\dfrac{\text{Arc length ABC}}{2\pi r} = \dfrac{\theta}{2\pi}$

Length of arc ABC $= \dfrac{\theta}{2\pi} \times 2\pi r = r\theta$

Example 4.4 A chord PQ of a circle of radius 5 cm subtends an angle of 50° at the centre. Find the length of the minor arc PQ. A chord PQ of a circle divides the circle in two parts. The larger arc is called the major arc of the circle and the smaller arc the minor arc.

$$\frac{\text{Arc length PQ}}{\text{Circumference of circle}} = \frac{\theta^\circ}{360^\circ}$$

$$\frac{\text{Arc length PQ}}{2\pi \times 5} = \frac{50}{360}$$

$$\text{Arc length PQ} = \frac{50}{360} \times 2\pi \times 5 = 4.36 \text{ cm}$$

or arc length PQ $= r\theta$

where $r = 5$ cm and $\theta = 50 \times \dfrac{\pi}{180} = 0.8727$ rad.

Arc length PQ $= 5 \times 0.8727 = 4.36$ cm.

Example 4.5 A belt passes over two pulleys as shown in fig. 4.2. If the pulleys are of radii 200 mm and 120 mm and are positioned with their centres 600 mm apart calculate the length of the belt, assuming there is no slack.

AB is a tangent to both circles, $\angle PAB = \angle ABQ = 90^\circ$
Draw RQ parallel to AB, then $\angle PRQ = 90^\circ$
In $\triangle PRQ$, PQ $= 600$ mm.
PR $=$ AP $-$ AR $=$ AP $-$ BQ $= 200 - 120 = 80$ mm
Using the theorem of Pythagoras RQ^2 = PQ^2 $-$ PR^2
RQ^2 $= 600^2 - 80^2 = 353\,600$
RQ $= \sqrt{353\,600} = 594.6$ mm

Also $\cos \alpha = \dfrac{\text{PR}}{\text{PQ}} = \dfrac{80}{600} = 0.1333 \quad \alpha = 82.34^\circ$

Reflex angle APD $= 360^\circ - 2\alpha = 360^\circ - 164.68^\circ$
$= 195.32^\circ$
Also angle BQC $= 164.68^\circ$ (since BQ is parallel to AP)

Arc length AED $= r\theta = 200 \times \left(195.32^\circ \times \dfrac{\pi}{180}\right)$

$= 681.8$ mm

Arc length BFC $= r\theta = 120 \times \left(164.68 \times \dfrac{\pi}{180}\right)$

$= 344.9$ mm

AB $=$ CD $=$ RQ $= 594.6$ mm
Total belt length $=$ AB $+$ CD $+$ arc AED $+$ arc BFC
$\qquad\qquad\qquad = 594.6 + 594.6 + 681.8 + 344.9$
$\qquad\qquad\qquad = 2215.9$ mm
The belt length $= 2216$ mm

EXERCISE 4.2

1. A chord AB of a circle of radius 60 mm subtends an angle of 70° at the centre. Find the length of the minor arc AB.

2. A chord PQ of a circle of radius 1.76 m subtends an angle of 85° at the centre. Find the length of the major arc PQ.

3. A belt wraps round a pulley of diameter 100 mm with a contact angle of 220°. Calculate the length of belt in contact with the pulley.

4. Two pulleys of diameters 400 mm and 600 mm are positioned with their centres 1000 mm apart. A belt passes over the two pulleys, as in fig. 4.2. Calculate the length of the belt required, assuming no slack.

5. Calculate the perimeter of the cam shown in fig. 4.3, all dimensions in mm.

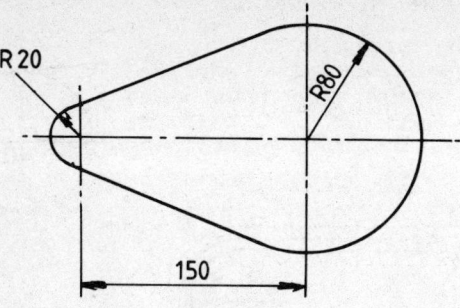

R 20

R80

150

Figure 4.3

4.3 SECTORS AND SEGMENTS

A sector of a circle is the part of a circle cut off by two radii, see fig. 4.4, OABC is a sector of the circle.

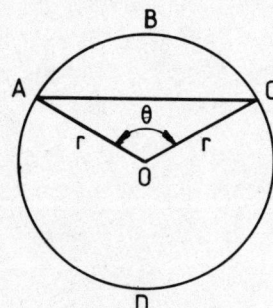

Figure 4.4

The area of the sector is proportional to the angle at the centre of the circle.

$$\frac{\text{Area of sector OABC}}{\text{Area of circle}} = \frac{\theta^\circ}{360^\circ} = \frac{\theta\,\text{rad}}{2\pi\,\text{rad}}$$

Working in radians,

$$\frac{\text{Area of sector OABC}}{\pi r^2} = \frac{\theta}{2\pi}$$

$$\text{Area of sector OABC} = \frac{\theta}{2\pi} \times \pi r^2 = \tfrac{1}{2} r^2 \theta$$

$$\text{Area of sector} = \tfrac{1}{2} r^2 \theta$$

A segment of a circle is the part of a circle cut off by a chord. In fig. 4.4, ABC is the minor segment and ADC is the major segment cut off by chord AB.
The area of segment ABC

$$\begin{aligned}
&= \text{area of sector OABC} - \text{area of } \triangle\text{OAC} \\
&= \tfrac{1}{2} r^2 \theta - \tfrac{1}{2} \text{OC} \times \text{OA} \times \sin\theta \\
&= \tfrac{1}{2} r^2 \theta - \tfrac{1}{2} r^2 \sin\theta \\
&= \tfrac{1}{2} r^2 (\theta - \sin\theta)
\end{aligned}$$

The area of $\triangle$OAC is found by using the formula for the area of a triangle: half the product of two sides times the sine of the angle between them. This is covered in Chapter 11, paragraph 11.2.

Example 4.6 A chord PQ of a circle of radius 90 mm subtends an angle of 64° at the centre O. Find the area of the sector POQ.

$$\frac{\text{Area of sector POQ}}{\text{Area of circle}} = \frac{64^\circ}{360^\circ}$$

$$\frac{\text{Area of sector POQ}}{\pi \times 90^2} = \frac{64}{360}$$

$$\text{Area of sector POQ} = \frac{64}{360} \times \pi \times 90^2 = 4524\,\text{mm}^2$$

or area of sector POQ $= \tfrac{1}{2} r^2 \theta$

Where $r = 90$ mm and $\theta = 64 \times \dfrac{\pi}{180} = 1.117$ rad

Area of sector POQ $= \tfrac{1}{2} \times 90^2 \times 1.117 = 4524\,\text{mm}^2$

Example 4.7 A circular metal shaft, of radius 50 mm has a flat of width 30 mm milled on it. Find the cross-sectional area of the shaft shown in fig. 4.5.

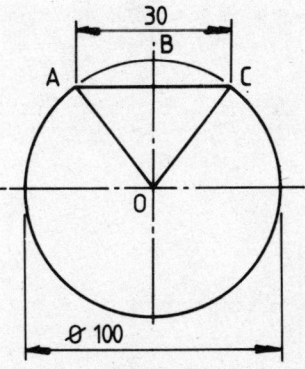

30

A · B · C

⌀ 100

Figure 4.5

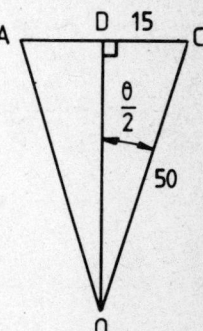

A   D 15   C

$\frac{\theta}{2}$

50

O

Figure 4.6

In fig. 4.5, area of segment ABC
= area of sector ABCO − area of triangle OAC

Area of segment $= \tfrac{1}{2} r^2 \theta - \tfrac{1}{2} r^2 \sin\theta$ (1)

To find the angle θ consider the triangle OAC, see fig. 4.6.

In $\triangle$ODC, OC = 50 mm, DC = 15 mm, $\angle$ODC = 90°

and $\angle\text{DOC} = \dfrac{\theta}{2}$

$\text{Sin} \angle \text{DOC} = \dfrac{\text{DC}}{\text{OC}}$

$\sin\dfrac{\theta}{2} = \dfrac{15}{50} = 0.3000$

$$\frac{\theta}{2} = 17°28'$$

or $\theta = 34°56'$ (0.6097 rad)

Substitute these values of θ into equation (1)

Area of segment ABC $= \frac{1}{2}r^2 (\theta - \sin \theta)$
$= \frac{1}{2} \times 50 \times 50 \times (0.6097 - 0.5726)$
$= 1250 \times 0.0371$
$= 46.4 \text{ mm}^2$

But the cross-sectional area of the shaft
= area of circle − area of segment

$= \pi \times 50 \times 50 - 46.4$
$= 7855 - 46.4$
$= 7809 \text{ mm}^2$

Cross-sectional area of the shaft = 7809 mm²

EXERCISE 4.3

1. A chord AB of a circle of radius 65 mm subtends an angle of 37° at the centre O. Find the area of the sector OAB.

2. A chord PQ of a circle of radius 2.73 m subtends an angle of 73° at the centre O. Find the area of the major sector OPQ.

3. A chord AB of a circle of radius 104 mm subtends an angle of 35° at the centre. Find the area of the minor segment cut off by the chord AB.

4. A chord AB of a circle, centre O and radius 86 mm, is 76 mm long. Find the area of the minor segment cut off by the chord AB.

5. A simple lift cam with straight sides has the dimensions shown in fig. 4.7 all dimensions in mm. Calculate the area of the view given.

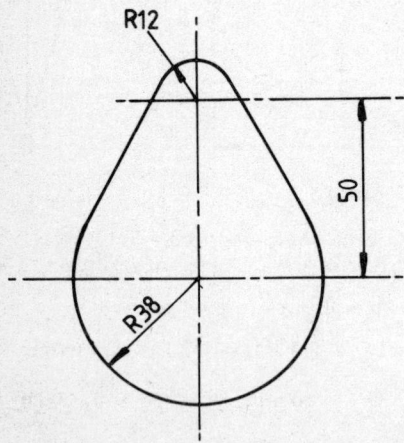

R12

50

R38

Figure 4.7

6. The radius of the arc of a circular segment is 60 mm and the angle it subtends at the centre is 35°. Calculate the area of the segment and the length of the chord.

4.4 GENERAL EXAMPLES

Example 4.8 The bracket shown in fig. 4.8, had to be made from 6 mm thick strip. Calculate the length of strip required at the mean thickness, i.e., the length of the centre line.

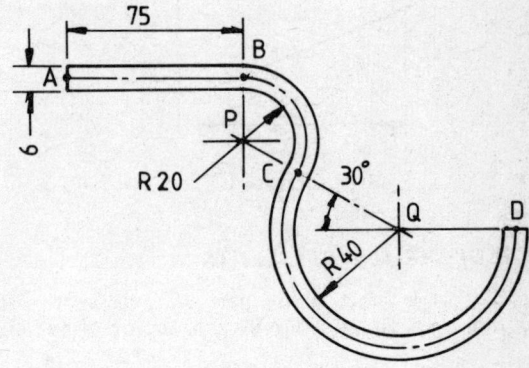

75

A   B

P

6

R20   C   30°

R40   Q   D

Figure 4.8

Length of strip = AB + arc BC + arc CD

For arc BC, radius = 20 + 3 = 23 mm, and the angle at the centre P is $90° + 30° = 120°$

$$\frac{\text{Arc BC}}{\text{Circumference}} = \frac{120°}{360°}$$

$$\text{Arc BC} = \frac{120}{360} \times (2\pi \times 23) = \frac{2\pi \times 23}{3}$$

Arc BC = 48.2 mm

For arc CD, radius = 40 + 3 = 43 mm, and the angle at the centre Q is $180° + 30° = 210°$

$$\frac{\text{Arc CD}}{\text{Circumference}} = \frac{210°}{360°}$$

$$\text{Arc CD} = \frac{210}{360} \times (2\pi \times 43) = \frac{7 \times 2\pi \times 43}{12}$$

Arc CD = 157.6 mm

Length of strip required = 75 + 48.2 + 157.6 = 281 mm

Length of strip required is 281 mm.

EXERCISE 4.4

1. The angle of a sector of radius 90 mm is 1.80 radians. Calculate the length of the arc and the area of the sector.

2. A sheet metal channel 1.2 m long has the cross-section shown in fig. 4.9. It is made from a blank whose width can be assumed to be equal to the length of the chain dotted centre line shown in the cross-section. Calculate the blank dimensions to three significant figures, all dimensions in mm.

3. Calculate the perimeter and the area shown of the cam drawn in fig. 4.10, all dimensions in mm.

4. Two pulleys of diameters 770 mm and 530 mm with centres 880 mm apart are connected by a cross belt, as shown in fig. 4.11. Calculate the total length of belt, assuming no slack.

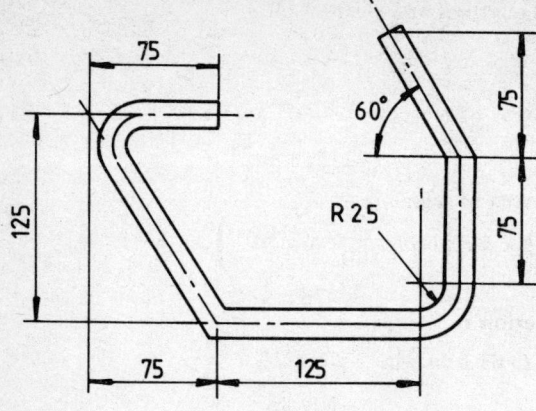

Figure 4.9

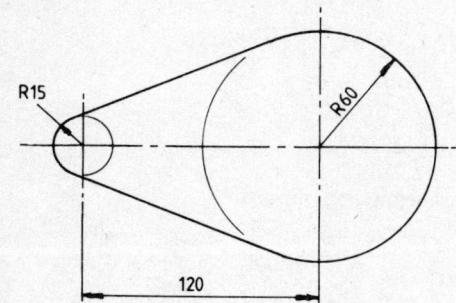

Figure 4.10

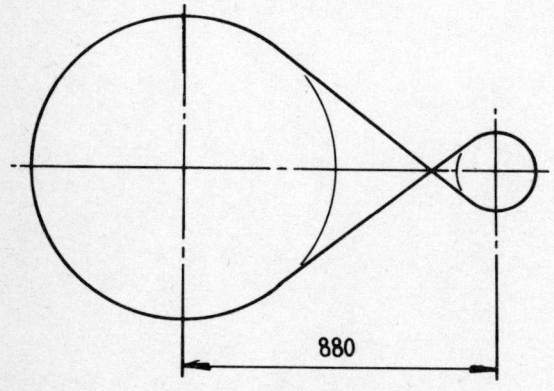

Figure 4.11

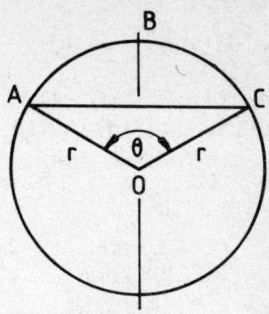

Figure 4.12

4. In fig. 4.12 the arc length $= 2\pi r \times \dfrac{\theta°}{360°}$ and the area of sector OABC $= \pi r^2 \times \dfrac{\theta°}{360°}$, where θ is in degrees.

SELF ASSESSMENT PAPER No 4

Instructions: Answer all questions in both sections
Time allowed: Section A 30 minutes (6 marks each question)
Section B 30 minutes (question 1, 12 marks; question 2, 18 marks)
Marks gained: 30+ pass with credit, 24−30 pass, less than 24 fail, repeat Chapter 4.

Section A

1. Write down the following angles in radians
(a) 45° (b) 87°23′ (c) 193°18′

2. Write down the following angles in degrees
(a) $\dfrac{\pi}{6}$ rad (b) 0.85 rad (c) 3.74 rad

3. Find the length of the arc of a circle of radius 100 mm if it subtends an angle of 60° at the centre of the circle.

4. Calculate the area of a sector of a circle of radius 60 mm if the angle at the centre of the circle is 75°.

5. A chord AB of a circle of radius 50 mm subtends an angle of 60° at the centre. Find the area of the minor segment cut off by the chord AB.

Section B

1. A piece of wire in the form of an arc of a circle of radius 90 mm and subtending an angle of 70° at the centre is bent into the form of a complete circle. Find its radius.

2. A circular steel shaft 1.25 m long and 45 mm diameter has a flat surface 20 mm wide milled along it from end to end. Calculate the volume of the finished shaft.

SUMMARY

1. A radian is the angle subtended at the centre of a circle of radius r by an arc of length r.
2. π radians $= 180°$.
3. In fig. 4.12 the length of arc ABC $= r\theta$, the area of the sector OABC $= \frac{1}{2}r^2\theta$ and the area of the segment ABC $= \frac{1}{2}r^2(\theta - \sin\theta)$. Where θ is in radians.

ANSWERS

Exercise 4.1   1. (a) $\dfrac{\pi}{2}$ (b) $\dfrac{\pi}{6}$ (c) $\dfrac{3\pi}{4}$

(d) 0.476   (e) 2.40   (f) 5.41   2. (a) $60°$
(b) $45°$ (c) $270°$ (d) $100°16'$
(e) $169°36'$ (f) $209°7'$ 3.   2.51 rev/s,
$901°50'$ 4. 100.5   5. 2.04 rev/s, 12.84 rad/s
6. (a) 64.9   (b) $3720°$

Exercise 4.2   1.   73.3 mm   2.   8.45 m
3.   192 mm   4.   3580 mm   5.   639 mm

Exercise 4.3   1.   1364 mm^2 2.   18.7 m^2
3.   202 mm^2 4.   453 mm^2 5.   5340 mm^2
6.   67.1 mm^2, 36.1 mm

Exercise 4.4   1. 162 mm,   7290 mm^2,
2.   1.20 m by 0.547 m   3.   493 mm, 15 640 mm
4.   3428 m

SELF ASSESSMENT PAPER No 4

Marks

Section A

1. (a) $\dfrac{\pi}{4}$ 2

(b) 1.525   2
(c) 3.374   2
2. (a) $30°$ 2
    (b) $48°42'$ 2
    (c) $214°17'$ 2

3. Length of arc $= 100 \times 60 \times \dfrac{\pi}{180}$ 3

$\qquad\qquad = 104.7$ mm   3

4. Area of sector $= \frac{1}{2} \times 60^2 \times 75 \times \dfrac{\pi}{180}$ 3

$\qquad\qquad = 2356$ mm^2 3

5. Area of segment $=$

$\frac{1}{2} \times 50^2\left(60 \times \dfrac{\pi}{180} - \sin 60°\right)$

$\qquad\qquad = 226$ m^2 3

Section B

1. Length of wire $= 90 \times 70 \times \dfrac{\pi}{180}$ 3

$\qquad\qquad = 109.96$ 2

$2\pi r = 109.96$ 4

$r = \dfrac{109.96}{2\pi}$ 2

$r = 17.5$ mm   1

2. Refer to fig. 4.6
$\sin \dfrac{\theta}{2} = \dfrac{10}{22.5}$ 2
$\theta = 52.78°$ 2
Area of segment removed
$= \frac{1}{2} \times 22.5^2 \left(52.78° \times \dfrac{\pi}{180} - \sin 52.78°\right)$ 4
$= 31.61$ 4
Area of shaft $= \pi \times 22.5^2 - 31.61 = 1558.8$ 3
Volume of shaft $= 1558.8 \times 1250$ mm^3 2
$\qquad\qquad = 0.001\,95$ m^2 1

5 Trigonometric Curves and the Sine Rule

5.1 GRAPHS OF TRIGNOMETRICAL FUNCTIONS

If we draw a right angled triangle with the hypotenuse equal to one then the opposite side has length sin θ and the adjacent side has length cos θ, see fig. 5.1. Therefore if we draw a circle of unit radius then the sine of the angle will always be the vertical projection of the radius, and the cosine will be the horizontal projection of the radius. For example in fig. 5.2, when $\theta = 120°$, sin $120° = 0.866$ and cos $120° = -0.5$.

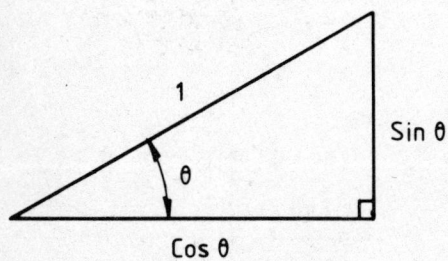

Figure 5.1

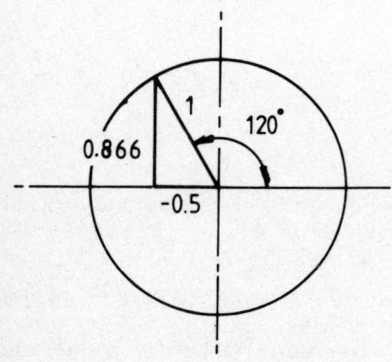

Figure 5.2

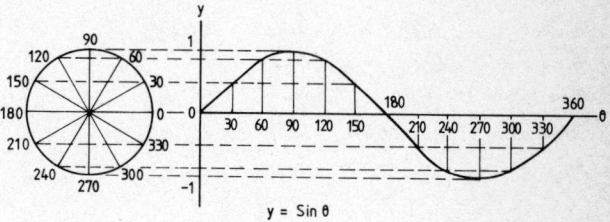

$y = \text{Sin } \theta$

Figure 5.3

We notice that all values of sin θ lie between $+1$ and -1. Also sin θ is positive for angles between $0°$ and $180°$ and negative for angles between $180°$ and $360°$. Similarly the cosine curve can be plotted, but with the cosine curve we must turn the circle through $90°$ so that the horizontal projections can be plotted on the graph. The graph for $y = \cos \theta$ is shown in fig. 5.4.

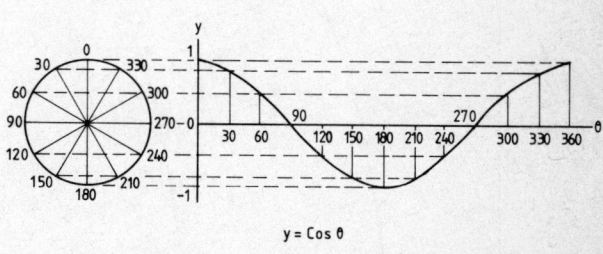

$y = \text{Cos } \theta$

Figure 5.4

If this process is used for angles from $0°$ to $360°$ then we obtain a complete cycle of the sine and cosine. For angles greater than $360°$ the process will repeat itself. Hence in fig. 5.3 we can see the graph of $y = \sin \theta$ for values of θ from $0°$ to $360°$.

It is seen that the graph for $y = \cos \theta$ is the same shape as the graph of $y = \sin \theta$ but that it is shifted through $90°$. Hence the values of cos θ lie between $+1$ and -1, also cos θ is positive for angles between $0°$ and $90°$ and $270°$ and $360°$ but is negative for angles between $90°$

and 270°. In level 1 we defined tan θ as the ratio $\dfrac{\text{opposite}}{\text{adjacent}}$.

Hence in fig. 5.1 $\tan \theta = \dfrac{\sin \theta}{\cos \theta}$

Using this definition we can draw up a table of values for tan θ, taking the values of sin θ and cos θ from the graphs in fig. 5.3 and fig. 5.4.

θ	0	30°	60°	90°	120°	150°	180°	210°
sin θ	0	0.5	0.866	1	0.866	0.5	0	-0.5
cos θ	1	0.866	0.5	0	-0.5	-0.866	-1	-0.866
tan θ	0	0.577	1.732	∞	-1.732	-0.577	0	0.577

240°	270°	300°	330°	360°
-0.866	-1	0.866	-0.5	0
-0.5	0	0.5	0.866	1
1.732	∞	-1.732	-0.577	0

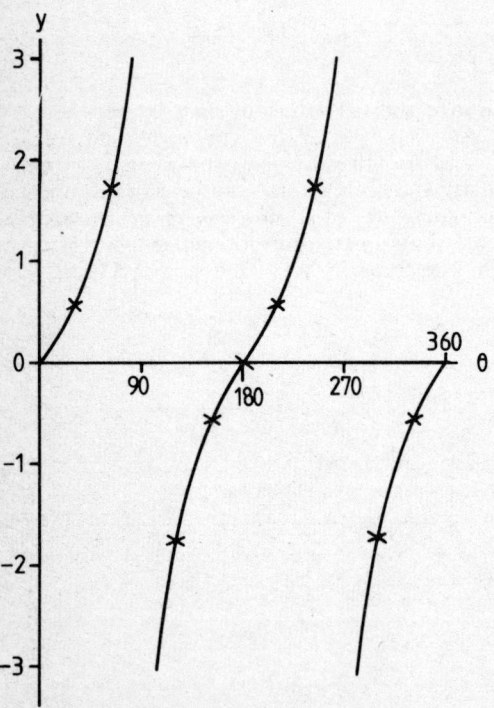

Figure 5.5

The graph of $y = \tan \theta$ is drawn in fig. 5.5. It can be seen that the values of tan θ are from $-\infty$ to $+\infty$ and that it repeats the values every 180°. Tan θ is positive from 0° to 90° and from 180° to 270°, it is negative from 90° to 180° and from 270° to 360°.

5.2 TRIGONOMETRICAL RATIOS FOR ANGLES BETWEEN 0° AND 360°

Previously, in level 1, we have found the sine, cosine and tangent of angles from 0° to 90°. These ratios for angles from 0° to 360° can be obtained directly from a calculator. When tables are used then the ratios are only given for angles from 0° to 90°. If, however, we look at the graphs in paragraph 5.1 we can see that:

(1) If the angle is between 90° and 180° subtract it from 180° and look up the required ratio in tables noting that the sine is positive, the cosine and the tangent negative.
(2) If the angle is between 180° and 270° subtract 180° from it and look up the required ratio in tables noting that the tangent is positive, the sine and cosine negative.
(3) If the angle is between 270° and 360° subtract it from 360° and look up the required ratio in tables noting that the cosine is positive, the sine and tangent negative. This is summarised in fig. 5.6.

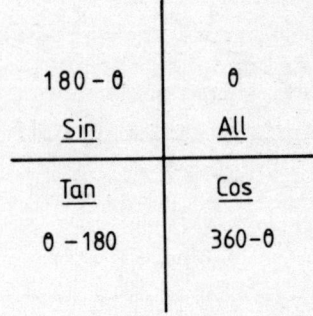

Figure 5.6

Example 5.1 Find the sine, cosine and tangent of the following angles (a) 37°44′,   (b) 138°10′, (c) 228°32′   (d) 305°26′

(a) Tables: sin 37°44′ = 0.6120, cos 37°44′ = 0.7909, tan 37°44′ = 0.7738.
Remember when using tables for cosines the mean difference column must be subtracted.
Calculator: The trigonometric values of 37°44′ can be found directly from a calculator, but on most calculators the minutes must first be changed to a decimal. CM 44 ÷ 60 + 37 = M+ F sin 0.6120 RM F cos 0.7909 RM F tan 0.7738

(b) Tables: 138°10′ is between 90° and 180°, we therefore subtract this from 180°. 180° − 138°10′ = 41°50′. We now look up the trigonometric ratios of 41°50′ and note that the cos and tan are negative.

$\sin 138°10' = \sin 41°50' = 0.6670$
$\cos 138°10' = -\cos 41°50' = -0.7451$
$\tan 138°10' = -\tan 41°50' = -0.8952$
Calculator: The trigonometric ratios can be found directly
CM $10 \div 60 + 138 = $ M+ F sin 0.6670 RM F cos -0.7451 RM F tan -0.8952

(c) Tables: $228°32'$ is between $180°$ and $270°$, we therefore subtract $180°$ from this. $228°32' - 180° = 48°32'$.
We now look up the trigonometric ratios of $48°32'$ and note that the sine and cosine are negative.
$\sin 228°32' = -\sin 48°32' = -0.7493$
$\cos 228°32' = -\cos 48°32' = -0.6622$
$\tan 228°32' = \tan 48°32' = 1.1316$
Calculator: The trignometric ratios can be found directly
CM $32 \div 60 + 228 = $ M+ F sin -0.7493 RM F cos -0.6622 RM F tan 1.1316

(d) Tables: $305°26'$ is between $270°$ and $360°$, we therefore subtract the angle from $360°$. $360° - 305°26' = 54°34'$. We therefore look up the trigonometric ratios of $54°34'$ and note that the sine and tangent are negative.
$\sin 305°26' = -\sin 54°34' = -0.8148$
$\cos 305°26' = \cos 54°34' = 0.5798$
$\tan 305°26' = -\tan 54°34' = -1.4054$
Calculator: The trigonometric ratios can be found directly
CM $26 \div 60 + 305 = $ M+ F sin -0.8148 RM F cos 0.5798 RM F tan -1.4054

Example 5.2 For the following equations find all of the solutions between $0°$ and $360°$.
(a) $\sin \theta = 0.5695$, (b) $\tan \theta = 1.2476$
(c) $\cos \theta = -0.6723$.

(a) Tables: Since the sine is positive the angle θ must be between $0°$ and $180°$.
Using tables, when $\sin \theta = 0.5695, \theta = 34°43'$
Hence the solutions are $\theta = 34°43'$ or $180° - 34°43'$
$\theta = 34°43'$ or $145°17'$
Calculator: 0.5695 F sin^{-1} 34.72° cs $+ 180 = 145.28°$
The first value is found from the calculator by using the inverse function key sin^{-1}. This answer must be subtracted from $180°$ to find the second value since the sine is positive between $0°$ and $180°$.
$\theta = 34.72°$ or $145.28°$

(b) Tables: Since the tan is positive the angle θ must be between $0°$ and $90°$ or between $180°$ and $270°$.
Using the tables when $\tan \theta = 1.2476, \theta = 51°17'$
Hence the solutions are $\theta = 51°17'$ or $180° + 51°17'$
$\theta = 51°17'$ or $231°17'$
Calculator: 1.2476 F tan^{-1} 51.29° $+ 180 = 231.29°$
The first value is found from the calculator by using the inverse function key tan^{-1}. This answer must be added to $180°$ to find the second value since the tangent is positive between $0°$ and $90°$ and between $180°$ and $270°$.
$\theta = 51.29°$ or $231.19°$

(c) Tables: Since the cosine is negative the angle θ must be between $90°$ and $270°$.
Using the tables when $\cos \theta = 0.6723, \theta = 47°45'$
Hence the solutions are $180° - 47°45'$ or $180° + 47°45'$
$\theta = 132°15'$ or $227°45'$
Calculator: -0.6723 F cos^{-1} 132.25° cs $+ 360 = 227.75°$
The first value is found from the calculator by using the inverse function key cos^{-1}. This answer must be subtracted from $360°$ to find the second value since the cosine is negative between $90°$ and $270°$. (The $132.25°$ is the $180 - \theta$ value, the $180 + \theta$ value is equal to $360 - (180 - \theta)$ hence the $132.25°$ is subtracted from 360.)
$\theta = 132.25°$ or $227.75°$
It will probably be easier to feed in 0.6723 to the calculator and then use the same method as that for tables. This is also true for the sine and tangent as when these are negative the calculator will give a negative angle and it is not always easy to get the required angles, between $0°$ and $360°$ from the calculator.

Example 5.3 Draw the graph of $y = \cos \theta$, for values of θ from $0°$ to $360°$. From your graph find (a) the value of $\cos \theta$ when $\theta = 289°$, (b) the value of θ when $\cos \theta = -0.75$

θ	0	30	60	90	120	150	180
$\cos \theta$	1	0.866	0.500	0	-0.5	-0.866	-1.0

θ	210		240	270	300	330	360
$\cos \theta$	-0.866		-0.5	0	0.5	0.866	1.0

The graph is drawn in fig. 5.7

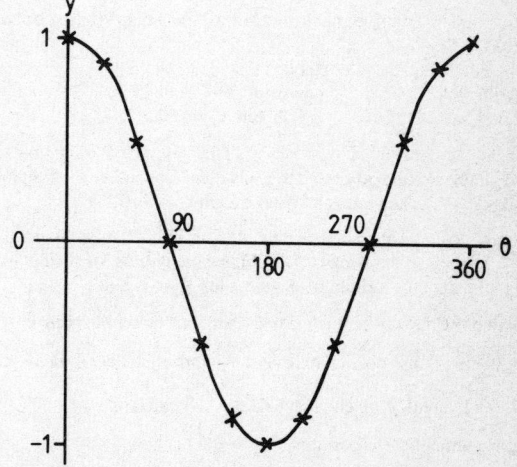

Figure 5.7

(a) When $\theta = 289°$, the value of $\cos \theta$ is 0.33
(b) When $\cos \theta = -0.75, \theta = 139°$ or $221°$

Example 5.4 The value of an alternating current I amps, in a circuit, after time t seconds is given by $I = 50 \sin 100\pi t$. Calculate,
(a) the time when the current first reaches 25 A,
(b) the value of the current after 0.015 s.

(a) $I = 50 \sin 100\pi t$
When $I = 25$, $25 = 50 \sin 100\pi t$

$$\sin 100\pi t = \frac{25}{50} = 0.5$$

$100\pi t$ is an angle in radians such that the sine of this angle is equal to 0.5. The angle is $30°$ or $\dfrac{\pi}{6}$ radians.

Therefore $100\pi t = \dfrac{\pi}{6}$

$$t = \frac{\frac{\pi}{6}}{100\pi} = \frac{\pi}{6} \times \frac{1}{100\pi} = \frac{1}{600} = 0.001\,67\,\text{s}$$

The current is 25 A when the time is 0.001 67 s

(b) $I = 50 \sin 100\pi t$
When $t = 0.015$, $I = 50 \sin 100\pi \times 0.015$
$I = 50 \sin 1.5\pi$

$$1.5\pi \text{ rad} = 1.5\pi \times \frac{180}{\pi} = 270°$$

$\sin 270° = -1$
$I = 50 \times -1 = -50\,\text{A}$

EXERCISE 5.2

1. Find the sine, cosine and tangent of the following angles
(a) $71°23'$,   (b) $153°49'$,   (c) $243°11'$
(d) $323°17'$ (e) $\dfrac{\pi}{3}$ rad,   (f) $\dfrac{4\pi}{3}$ rad.

2. For the following equations find all of the solutions between $0°$ and $360°$.
(a) $\sin \theta = 0.8742$,   (b) $\cos \theta = 0.0924$,
(c) $\tan \theta = 1.2742$ (d) $\sin \theta = -0.2431$,
(e) $\cos \theta = -0.5762$,   (f) $\tan \theta = -0.8724$

3. Draw the graph of $y = \sin \theta$, for values of θ from $0°$ to $360°$. From your graph, find, (a) the value of $\sin \theta$ when $\theta = 160°$, (b) the values of θ when $\sin \theta = -0.75$.

4. Draw the graph of $y = \tan \theta$, for values of θ from $0°$ to $360°$. From your graph, find, (a) the values of $\tan \theta$ when $\theta = 160°$, (b) the values of θ when $\tan \theta = 1.27$.

5. Draw the graph of $y = \cos \theta$, for values of θ from 0 to 2π radians, plotting points at every $\dfrac{\pi}{6}$ radians. From your graph, find, (a) the value of $\cos \theta$ when $\theta = \dfrac{5\pi}{12}$ radians
(b) the values of θ when $\cos \theta = -0.71$.

6. Evaluate $\cos 2\pi ft$, if $f = 50$ Hz and $t = 0.006$ s.

7. Evaluate $\sin wt$, if $w = 40$ rad/s and $t = 0.1$ s.

8. The value of an alternating current, I amps, in a circuit, after time t seconds is given by, $I = 100 \sin 50\pi t$.
Calculate, (a) the time when the current first reaches 50 A, (b) the value of the current after 0.015 s.

5.3 PERIODIC PROPERTIES AND TRIGONOMETRIC IDENTITIES

In paragraph 5.2 we looked at the trigonometric ratios of angles from $0°$ to $360°$. If we consider the way we drew the graphs from a rotating vector in paragraph 5.1 we can see that as the angle increases above $360°$ so the graph repeats itself. Similarly, if the rotating vector is rotated in the opposite direction, the graph again repeats itself. Thus the graph for angles from $0°$ to $360°$ is just part of the complete graph which will be continuous in both directions. We say that the trigonometric functions $\sin \theta$ and $\cos \theta$ are periodic and have a period of $360°$ or 2π radians. We will see later that if we change the angle to 2θ then as θ goes from 0 to $180°$, $\sin 2\theta$ goes through one complete cycle, or the period is halved.

It will be noticed that whatever the value of the angle the value of $\sin \theta$ and $\cos \theta$ will never be less than -1 nor greater than $+1$. We call this the amplitude. The amplitude of $\sin \theta$ and $\cos \theta$ is 1. To change the amplitude we must multiply $\sin \theta$ by a constant. Thus $a \sin \theta$ has an amplitude of a.

If we now look at the graph of $\tan \theta$, see fig. 5.5, this is also periodic but its period is $180°$. That is it repeats its values every $180°$.

When we first considered $\tan \theta$ in this chapter we looked at a right angled triangle where the hypotenuse was 1, see fig. 5.1.

Here $\tan \theta = \dfrac{\sin \theta}{\cos \theta}$. This is called a trigonometric identity since it is true for all values of θ. Similarly if we apply Pythagoras' theorem to this triangle we have

$$(\sin \theta)^2 + (\cos \theta)^2 = 1$$

This has been derived from a right angled triangle but is true for all values of θ.
It is usual to remove the brackets in $(\sin \theta)^2$ and write it as $\sin^2 \theta$. But one must always remember that it is really $(\sin \theta)^2$.
Thus $\sin^2 \theta + \cos^2 \theta = 1$ for all values of θ.

Example 5.5 Find the sine, cosine and tangent of
(a) $797°$ (b) 11π rad

(a) $797° = 360° + 360° + 77°$. Hence the trigonometric ratios of $797°$ will be the same as those for $77°$
$\sin 797° = \sin 77° = 0.9744$
$\cos 797° = \cos 77° = 0.2250$
$\tan 797° = \tan 77° = 4.3315$

(b) 11π rad $= 2\pi + 2\pi + 2\pi + 2\pi + 2\pi + \pi$. Hence the trigonometric ratios of 11π will be the same as those for π rad.
$\sin 11\pi = \sin \pi = 0$
$\cos 11\pi = \cos \pi = -1$
$\tan 11\pi = \tan \pi = 0$

Example 5.6 Evaluate $\cos 2\pi ft$ if $f = 100$ Hz and $t = 0.024$ s

$$\cos 2\pi ft = \cos(2\pi \times 100 \times 0.024)$$
$$= \cos 4.8\pi$$

$4.8\pi = 2\pi + 2\pi + 0.8\pi$, hence $\cos 4.8\pi$ is the same as $\cos 0.8\pi$.

$$\cos 2\pi ft = \cos 4.8\pi = \cos 0.8\pi = \cos 144°$$
$$= \cos(180° - 144°)$$
$$= \cos 36°$$
$$= 0.809$$

Example 5.7 Prove that $\cos\theta - \sin\theta = \dfrac{1 - 2\sin^2\theta}{\cos\theta + \sin\theta}$

We will start with the right hand side (RHS) of this identity.

$$\frac{1 - 2\sin^2\theta}{\cos\theta + \sin\theta} = \frac{\cos^2\theta + \sin^2\theta - 2\sin^2\theta}{\cos\theta + \sin\theta}$$

writing $1 = \cos^2\theta + \sin^2\theta$

$$= \frac{\cos^2\theta - \sin^2\theta}{\cos\theta + \sin\theta} = \frac{(\cos\theta + \sin\theta)(\cos\theta - \sin\theta)}{\cos\theta + \sin\theta}$$

Dividing top and bottom by $\cos\theta + \sin\theta$
$= \cos\theta - \sin\theta =$ left hand side (LHS).

EXERCISE 5.3

1. Find the sine, cosine and tangent of (a) 879°, (b) 682°, (c) 1392°, (d) 8π rad, (e) $\dfrac{15\pi}{2}$ rad

2. Evaluate $\cos 2\pi ft$, if $f = 200$ Hz and $t = 0.024$ s.

3. Evaluate $\sin wt$, if $w = 50$ rad/s and $t = 0.3$ s.

4. Prove that $(1 + \tan^2\theta)\cos^2\theta = 1$ for all values of θ.

5. Prove that $1 - 2\sin^2\theta = 2\cos^2\theta - 1$.

5.4 THE SINE RULE

The normal notation used is that in any triangle ABC the sides opposite the angles A, B and C are of lengths a, b and c respectively. The angles A, B and C refer to the interior angles of the triangle ABC.
We consider two triangles as shown in fig. 5.8

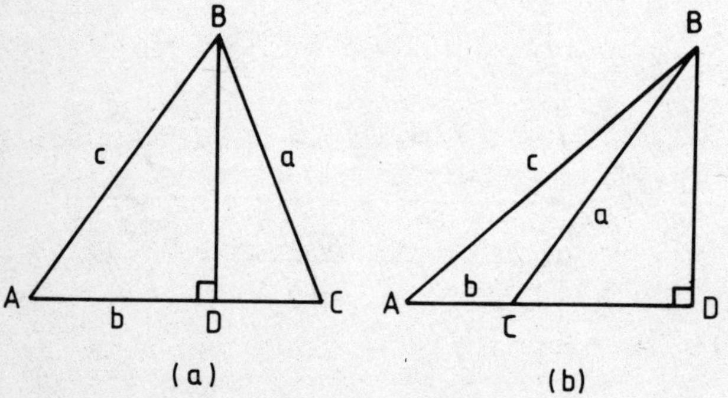

In fig. 5.8(a)
In $\triangle ABD$, $\angle ADB = 90°$

$$\sin A = \frac{BD}{AB} = \frac{BD}{c}$$

$BD = c\sin A \qquad (1)$
In $\triangle BDC$, $\angle BDC = 90°$

$$\sin C = \frac{BD}{BC} = \frac{BD}{a}$$

$BD = a\sin C \qquad (2)$

In fig. 5.8(b)
In $\triangle ABD$, $\angle ABD = 90°$

$$\sin A = \frac{BD}{AB} = \frac{BD}{c}$$

$BD = c\sin A \qquad (1)$
In $\triangle BDC$, $\angle BDC = 90°$

$$\sin\angle BCD = \frac{BD}{BC} = \frac{BD}{a}$$

But
$$\sin\angle BCD = \sin(180° - C)$$
$$= \sin C$$
$$\sin C = \frac{BD}{a}$$

or $BD = a\sin C \qquad (2)$

From (1) and (2)
$a\sin C = c\sin A$

or $\dfrac{a}{\sin A} = \dfrac{c}{\sin C}$

In both figures, if the perpendicular from C to AB is drawn it can be shown that $\dfrac{a}{\sin A} = \dfrac{b}{\sin B}$

Hence in any triangle $\dfrac{a}{\sin A} = \dfrac{b}{\sin B} = \dfrac{c}{\sin C}$

The sine rule is used to solve a triangle when either
(a) two angles and one side are given
or (b) two sides and an angle opposite one of these sides is given.

Example 5.8 Find a and b in the triangle ABC if $A = 60°$, $C = 73°$ and $c = 15$ m.

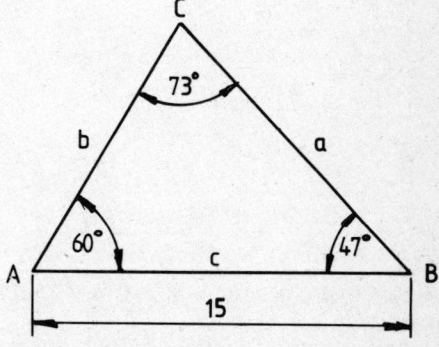

Figure 5.9

$B = 180° - 60° - 73° = 47°$
Using the sine rule

$$\frac{a}{\sin 60°} = \frac{b}{\sin 47°} = \frac{15}{\sin 73°}$$

$$a = \frac{15 \times \sin 60°}{\sin 73°} = \frac{15 \times 0.8660}{0.9563} = 13.6 \text{ m}$$

Figure 5.8

(a)   (b)

similarly $b = \dfrac{15 \times \sin 47°}{\sin 73°} = \dfrac{15 \times 0.7314}{0.9563} = 11.5$

Or using a calculator for the sines

CM 73 F sin M+ 47 sin × 15 ÷ RM = 11.5

$a = 13.6$ m, $b = 11.5$ m.

Example 5.9 Calculate the length x of the template shown in fig. 5.10, all dimensions in mm.

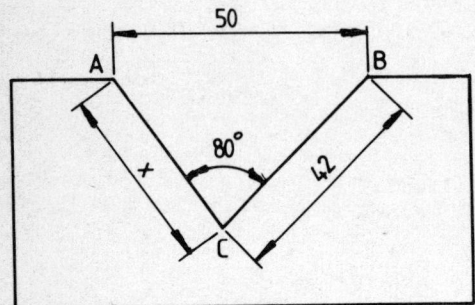

Figure 5.10

Using the sine rule on triangle ABC

$\dfrac{a}{\sin A} = \dfrac{b}{\sin B} = \dfrac{c}{\sin C}$

$\dfrac{42}{\sin A} = \dfrac{x}{\sin B} = \dfrac{50}{\sin 80°}$

$\dfrac{42}{\sin A} = \dfrac{50}{\sin 80°}$ or

$\sin A = \dfrac{42 \times \sin 80°}{50}$

Using a calculator

80 F sin × 42 ÷ 50 = F sin^{-1} 55.82°

$A = 55.82°$
But $B = 180° - 55.82° = 44.18°$

$\dfrac{x}{\sin 44.18°} = \dfrac{50}{\sin 80°}$ $x = \dfrac{50 \times \sin 44.18°}{\sin 80°}$

CM 80 F sin M+ 44.18 F sin × 50 ÷ RM = 35.4
The dimension x is 35.4 mm.
In this example we took the angle A to be 55.82° because sin A = 0.8272. But when sin A = 0.8272 then there are two possible answers. 55.82° or 180 − 55.82° = 124.18°. In this case the answer cannot be 124.18° since one angle of the triangle is 80° and 80° + 124.18° is greater than 180°. However, we must always check to see that we have the correct angle, or in some cases there are two possible answers. If there are two possible answers then this is called the ambiguous case.

Example 5.10 Three rods are joined to form a framework as shown in fig. 5.11. Calculate the angle C.

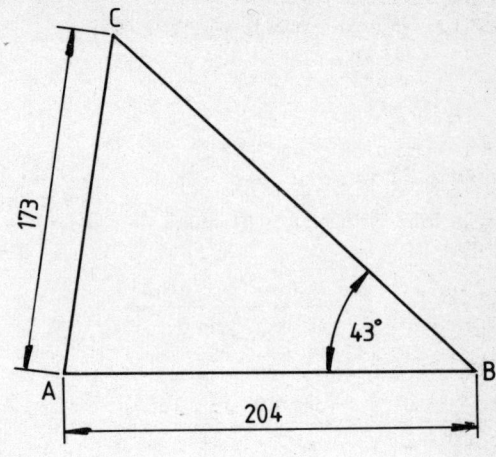

Figure 5.11

$\dfrac{a}{\sin A} = \dfrac{b}{\sin B} = \dfrac{c}{\sin C}$

$\dfrac{a}{\sin A} = \dfrac{173}{\sin 43°} = \dfrac{204}{\sin C}$

or $\sin C = \dfrac{204 \times \sin 43°}{173} = 0.8042$

$C = 53.5°$ or $126.5°$
In this case angle C could be either 53.5° or 126.5° and without further information we cannot say which is correct.

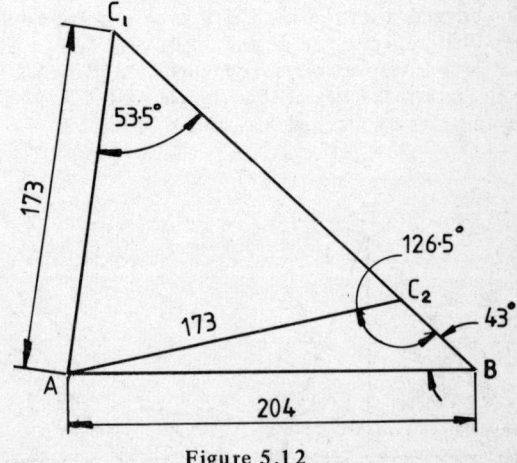

Figure 5.12

The two possible triangles are shown in fig. 5.12, they are ABC$_1$ and ABC$_2$

EXERCISE 5.4

1. Find a and b in the triangle ABC if $A = 37°$, $C = 80°$ and $c = 190$ mm.

2. Find angle A in the triangle ABC if $B = 62°$, $a = 52$ mm and $b = 64$ mm.

3. Find angle Q in the triangle PQR if $q = 33$ mm, $r = 39$ mm and $R = 74°$.

4. Find x and y in the triangle XYZ if $X = 77°$, $Z = 56°$ and $z = 3.6$ m.
5. Find the two values for R in the triangle PQR if $P = 27°$, $p = 460$ mm and $r = 610$ mm.

6. The jib AB of a crane is 14 m long, see fig. 5.13, and makes an angle of 112° with the horizontal base CA. Calculate the length of the stay CB if $\angle BCA = 31°$.

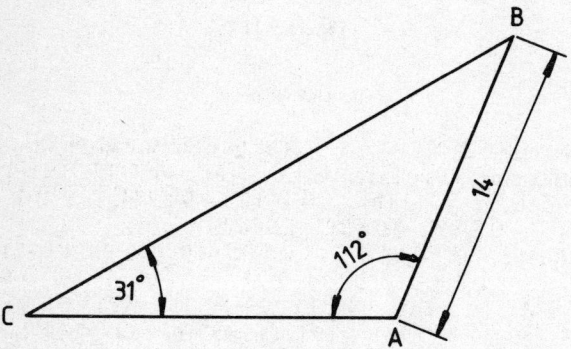

Figure 5.13

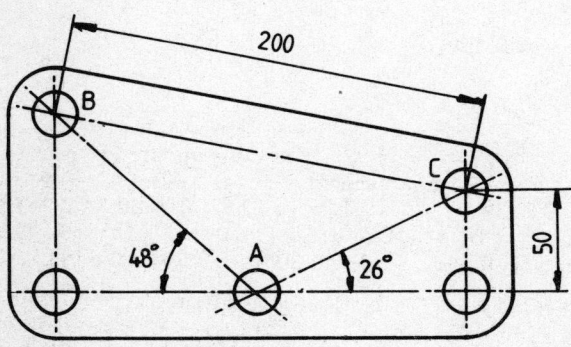

Figure 5.14

7. Fig. 5.14 shows a cover plate, calculate: (i) length AC, (ii) angle ABC, (iii) length AB.

8 AB and BC are vectors representing the alternating currents in two branches of a parallel circuit. The line AC is the resultant. If AB is equivalent to 40 A, BC is equivalent to 30 A, and $\angle BCA = 30°$, calculate the resultant current AC and the angle it makes with AB.

5.5 GENERAL EXAMPLES

Example 5.11 The value of a current is given by $I = 50 \sin (300t - 0.2)$. Calculate (a) I when $t = 0.004$ s, (b) the maximum value of I and the value of t when this first occurs.

(a) When $t = 0.004$, $I = 50 \sin (300 \times 0.004 - 0.2)$
$$I = 50 \sin 1.0$$

Now 1.0 rad $= 1.0 \times \dfrac{180}{\pi} = 57.30°$

Hence $I = 50 \times \sin 57.30° = 42.08$
The current is 42.08 A when $t = 0.004$ s.

(b) The maximum value of I will occur when the sine is a maximum. Now the sine cannot be greater than $+1$.
Hence the maximum value of I is $I = 50 \times 1 = 50$ A.
This will occur when $\sin (300t - 0.2) = 1$

$$\text{or } 300t - 0.2 = \frac{\pi}{2} \quad \left(\frac{\pi}{2} = 90°\right)$$

$$300t = \frac{\pi}{2} + 0.2$$

$$300t = 1.771$$
$$t = 0.0059 \text{ s}$$

The maximum current is 50 A and first occurs at 0.0059 s.

EXERCISE 5.5

1. The value of an alternating current, I amps, in a circuit, after time t seconds is given by, $I = 90 \sin 120\pi t$.
Calculate, (a) the time when the current first reaches 45 A, (b) the value of the current after 0.045 s.

2. If $I = 150 \sin \left(80\pi t - \dfrac{\pi}{4}\right)$ calculate, (a) the value of I when $t = 0$,   (b) the smallest positive value of t when $I = 0$.

3. A jib crane standing on horizontal ground has a vertical post AB 3.2 m long, a jib AC 6.3 m long and a tie BC. If $\angle ABC = 115°$ calculate   (a) the length of the tie BC, (b) the height of C above the level ground.
4. A crank OP of length 30 mm rotates about 0, see fig. 5.15. The connecting rod PQ is 84 mm long and Q moves in a slide AB. Calculate the two lengths OQ when $\angle PQO = 15°$.

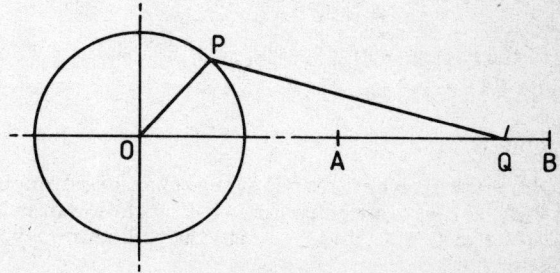

Figure 5.15

SUMMARY

1. The sine and cosine curves are periodic with a period of $360°$ or 2π radians. The sines and cosines have values between -1 and $+1$.

2. The tangent curve has a period $180°$ or π radians. The tangent can take all values.

3.

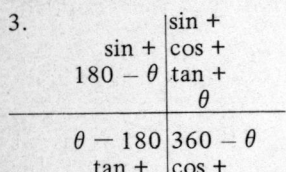

4. For all values of θ, $\tan \theta = \dfrac{\sin \theta}{\cos \theta}$ and $\sin^2\theta + \cos^2\theta = 1$.

5. In any triangle ABC $\dfrac{a}{\sin A} = \dfrac{b}{\sin B} = \dfrac{c}{\sin C}$, this is the sine rule.

6. The ambiguous case occurs when the sine rule gives two possible solutions.

SELF ASSESSMENT PAPER No 5

Instructions: Answer all questions in both sections
Time allowed: Section A 45 minutes (12 marks each question)
Section B 30 minutes (16 marks each question)
Marks gained: 40+ pass with credit, 32–40 pass, less than 32 fail, repeat chapter 5.

Section A

1. Find the following: (a) $\sin 69°17'$, (b) $\cos 235°14'$ (c) $\tan 154°39'$,   (d) $\sin 322°56'$.

2. Solve the following equations giving all solutions between $0°$ and $360°$ (a) $\sin \theta = -0.7438$, (b) $\cos \theta = 0.2714$,   (c) $\tan \theta = 1.8561$.

3. Prove the following identities (a) $\dfrac{\sin^2 A}{1 - \cos A} = 1 + \cos A$

(b) $\tan A + \dfrac{1}{\tan A} = \dfrac{1}{\sin A \cos A}$

4. In the triangle ABC, A = $79°$, C = $56°$ and $c = 22$ m. Calculate a and b.

Section B

1. Draw the graph of $y = \cos \theta$, for values of θ from $0°$ to $360°$. From your graph find   (a) the values of $\cos \theta$ when $\theta = 127°$,   (b) the values of θ when $\cos \theta = 0.65$.

2. In the layout shown in fig. 5.16 calculate the co-ordinate dimensions x and y.

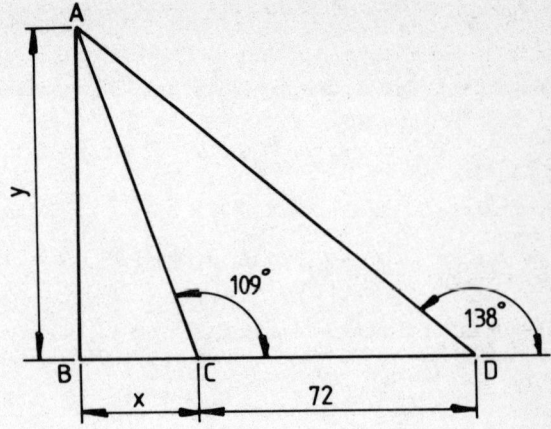

Figure 5.16

ANSWERS

Exercise 5.2   1.   The answers are given in the order sine, cosine, tangent,   (a) 0.9477, 0.3192, 2.9686,   (b)   0.4412, $-$ 0.8974, $-$ 0.4917, (c) $-$ 0.8925, $-$ 0.4511, 1.9782,   (d) $-$ 0.5979, 0.8016, $-$ 0.7458,   (e) 0.8660, 0.5000, 1.7321, (f) $-$ 0.8660, $-$ 0.5, 1.7321,   2.(a) $60°57'$, $119°3'$,   (b) $84°42'$, $275°18'$,   (c) $51°52'$, $231°52'$,   (d) $194°4'$, $345°56'$, (e) $125°11'$, $234°49'$ (f) $138°54'$, $318°54'$,   3.(a)   0.34, (b) $228°$, $312°$,   4.(a) $-$ 0.36,   (b) $52°$, $232°$,   5.(a)   0.26,   (b) $\dfrac{3\pi}{4}, \dfrac{5\pi}{4}$,

6. $-$ 0.309,   7. $-$ 0.7568,   8. (a) $\dfrac{1}{300}$ s,

(b) 70.7 A.

Exercise 5.3   1.   The answers are given in the order sine, cosine, tangent,   (a)   0.3584, $-$0.9336, $-$ 0.3839,   (b) $-$ 0.6157, 0.7880, $-$ 0.7813, (c) $-$ 0.7431, 0.6691, $-$ 1.1106,   (d)   0, 1, 0, (e)   1, 0, ∞,   2.   0.3090,   3.   0.6503.

Exercise 5.4   1. $a = 116$ mm $b = 172$ mm, 2. $45°50'$,   3. $54°26'$,   4. $x = 4.23$ m, $y = 3.18$ m,   5. $37°1'$, $142°59'$,   6.   25.2 m, 7.   (i) 114 mm, (ii)136 mm, (ii) $33°10'$,   8.   63 A, $22°1'$.

Exercise 5.5   1. (a) 0.001 39 s,   (b) $-$ 85.6 A, 2.(a)   106 A,   (b)   0.0031 s,   3.(a)   4.24 m, (b)   5.0 m,   4.   60.5 mm, 101.8 mm.

SELF ASSESSMENT PAPER No 5

Marks

Section A
1. (a) 0.9353   3

	Marks
(b) $\cos 235°14' = -\cos 55°14' = -0.5702$	3
(c) $\tan 154°39' = -\tan 25°21' = -0.4738$	3
(d) $\sin 322°56' = -\sin 37°4' = -0.6027$	3
2(a) $228°3', 311°57'$	4
(b) $74°15', 285°45'$	4
(c) $61°41', 241°41'$	4

3. (a) LHS $= \dfrac{1 - \cos^2 A}{1 - \cos A}$ 4

$\qquad\quad = \dfrac{(1 - \cos A)(1 + \cos A)}{1 - \cos A}$ 4

$\qquad\quad = 1 + \cos A = \text{RHS}$ 4

(b)   LHS $= \dfrac{\sin A}{\cos A} + \dfrac{\cos A}{\sin A}$ 4

$\qquad\quad = \dfrac{\sin^2 A + \cos^2 A}{\sin A \cos A}$ 4

$\qquad\quad = \dfrac{1}{\sin A \cos A} = \text{RHS}$ 4

4. $\dfrac{a}{\sin 79°} = \dfrac{b}{\sin 45°} = \dfrac{22}{\sin 56°}$ 4

$a = \dfrac{22 \sin 79°}{\sin 56°}$ 2

$a = 26.0$ m   2

$b = \dfrac{22 \sin 45°}{\sin 56°}$ 2

$b = 18.8$ m   2

Section B
1. Graph (see fig. 5.7)   10
(a) -0.60 2
(b) $49°, 311°$ 4

2. $\angle ADC = 42°, \angle CAD = 29°$ 2

$AC = \dfrac{72 \sin 42°}{\sin 29°}$ 2

$AC = 99.4$ mm   2

$\angle ACB = 71°$ 2
$x = 99.4 \cos 71°$ 2
$x = 32.4$ mm   2
$y = 99.4 \sin 71°$ 2
$y = 94.0$ mm   2

PHASE TEST 2

Instructions: Answer all questions
Time allowed: 1 hour
Marks gained: 30+ pass with credit, 24–30 pass, less than 24 fail, repeat, chapters 4 and 5.

1. Change the following angles in degrees to radians (i) $127.63°$, (ii) $58°37'$.

2. Change the following angles in radians to degrees (i) $\dfrac{3\pi}{4}$ rad, (ii) 2.46 rad.

3. Find the length of the arc of a circle of radius 75 mm if it subtends an angle of $87°$ at the centre.

4. Calculate the area of the sector formed by the arc in question 3.

5. Find the sine, cosine and tangent of the following angles (a) $77.84°$, (b) $235°$

6. Show that $\cos^2 \theta - \sin^2 \theta = 1 - 2 \sin^2 \theta$.

7. In triangle ABC, $a = 1.7$ m, $b = 2.3$ m and $A = 39°$. Calculate the angle B.

8. A belt wraps round a pulley of diameter 80 mm with a contact angle of $210°$. Calculate the length of belt in contact with the pulley.

9. Two vector quantities are represented by the sides AB and BC of a triangle ABC. The resultant is represented by AC. If AB $= 17.2$, $\angle ABC = 137°$ and AC $= 28.9$ find the magnitude of BC.

ANSWERS PHASE TEST 2

	Marks
1. (i) 2.228 radians, (ii) 1.023 radians	4
2. (i) $135°$, (ii) $140.95°$ $(140°57')$	4

3. Arc length $= 75 \times 87 \times \dfrac{\pi}{180}$

or $2\pi \times 75 \times \dfrac{87}{360}$ 4

$\quad = 113.9$ mm   2

4. Area of sector $= \frac{1}{2} \times 75^2 \times 87 \times \dfrac{\pi}{180}$

or $\pi \times 75^2 \times \dfrac{87}{180}$ 4

$\quad = 4271$ mm^2 2

5. (a) $\sin 77.84° = 0.9776$, $\cos 77.84° = 0.2106$
$\quad \tan 77.84° = 4.641$ 3

(b) $\sin 235° = -\sin 55° = -0.8192$ 2
$\quad \cos 235° = -\cos 55° = -0.5736$ 2
$\quad \tan 235° = \tan 55° = 1.4281$ 2

	Marks			Marks

6. R.H.S. $1 - 2\sin^2\theta = (1 - \sin^2\theta) - \sin^2\theta$ 2
$= \cos^2\theta - \sin^2\theta$ 2

$\sin C = \dfrac{17.2 \times \sin 137^\circ}{28.9}$ 2

7. $\dfrac{1.7}{\sin 39^\circ} = \dfrac{2.3}{\sin B}$ 3

$C = 23.95^\circ \ (23^\circ 57')$ 2
$A = 180^\circ - (23.95^\circ + 137^\circ) = 19.05^\circ$ 1

$\sin B = \dfrac{2.3 \times \sin 39^\circ}{1.7}$. 2

$\dfrac{BC}{\sin 19.05^\circ} = \dfrac{28.9}{\sin 137^\circ}$ 2

$B = 58.37^\circ \ (58^\circ 22')$ 2

$BC = \dfrac{28.9 \times \sin 19.05^\circ}{\sin 137^\circ}$ 2

8. Length of arc $= 40 \times 210 \times \dfrac{\pi}{180}$ 4
$\qquad\qquad = 146.6$ mm   2

$BC = 13.8$ 2

9. $\dfrac{28.9}{\sin 137^\circ} = \dfrac{17.2}{\sin C}$ 3

6 Measures of Central Tendency

6.1 THE MODE

The mode of a set of variables is the value of the variable which occurs most frequently.

Example 6.1 Find the mode of the following sets of variables

(a)   2, 3, 3, 5, 6, 7, 7, 7, 8, 8, 9

(b)   5, 5, 6, 6, 6, 7, 7, 8, 8, 8, 9, 9, 10

(c)   9. 8, 8, 7, 1, 2, 5, 4, 5, 6, 9, 5, 3, 6

In (a) the mode is 7 since this occurs three times, that is the most often.
In (b) there are two modes 6 and 8 since they both occur three times, which is the most often.
In (c) the numbers must be grouped   1, 2, 3, 4, 5, 5, 5, 6, 6, 7, 8, 8, 9, 9. The mode is 5.

When we have a frequency distribution the modal group is the largest group but the mode is still the value of the variable which occurs most often. If the mode is required it can be estimated from a histogram with regular class intervals.

Example 6.2 The diameters of 90 ball bearings were measured and the results recorded in the frequency distribution below.

Diameter d (mm)	6.55–6.56	6.56–6.57	6.57–6.58
Frequency f	4	12	22

6.58–6.59	6.59–6.60	6.60–6.61	6.61–6.62
19	16	11	6

(a) Which is the modal class?
(b) Draw a histogram and estimate the value of the mode.
(a) The modal class is the 6.57–6.58 class.
(b) The histogram is drawn in fig. 6.1.

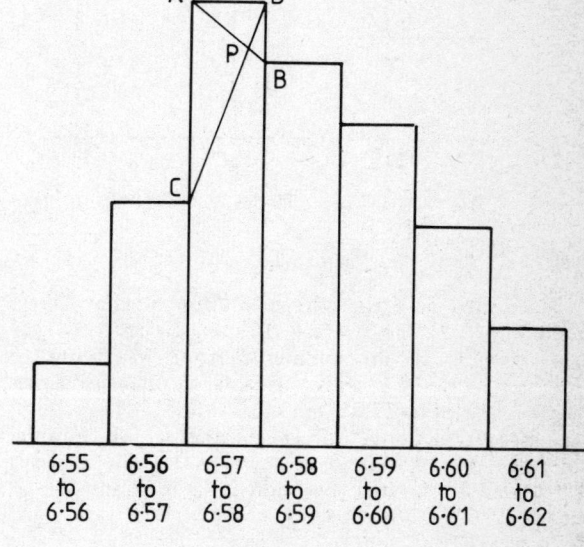

6·55 to 6·56	6·56 to 6·57	6·57 to 6·58	6·58 to 6·59	6·59 to 6·60	6·60 to 6·61	6·61 to 6·62

Figure 6.1

The value of the mode is found by drawing the two lines AB, CD and reading off the value of the variable at the point of intersection. The mode is 6.578 mm.

Example 6.3 The lengths of 100 components were recorded in the frequency table below.

Length (mm)	17.7	17.8	17.9	18.0	18.1	18.2	18.3	18.4
Frequency	3	11	17	27	24	11	5	2

(a) Which is the modal class?
(b) Find the value of the mode.

(a) The modal class is the fourth class, 18.0 mm
(b) It is not necessary to draw the whole histogram, it is just necessary to draw the middle part as shown in fig. 6.2.

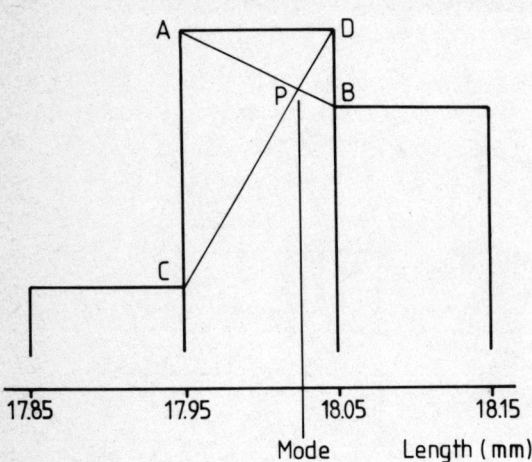

Figure 6.2

It is assumed here that the measurements have been rounded off to the nearest 0.1 mm. Hence the 17.7 class contains all measurements from 17.65 mm to 17.75 mm, the 17.8 class contains all measurements from 17.75 mm to 17.85 mm etc.

The measurement at the beginning of the modal group will be 17.95 and the end 18.05. The mode is determined by finding the point P, as in example 6.2, and is 17.95 + 0.08.

The mode is 18.03 mm.

If required the mode can be calculated from fig. 6.2

$$\text{Mode} = \text{value at beginning of modal group} + \left(\frac{AC}{AC + DB}\right)$$

$$\times \text{ width of modal group.}$$

$$\text{Mode} = 17.95 + \left(\frac{10}{10 + 3}\right) \times 0.1$$

$$\text{Mode} = 17.95 + 0.077 = 18.03 \text{ mm}$$

EXERCISE 6.1

1. Find the mode of the following:
(a) 6, 7, 7, 9, 11, 11, 11, 12, 14, 14, 15
(b) 3, 3, 4, 4, 5, 5, 5, 5, 7, 7, 8, 8, 8, 8, 9, 10
(c) 4, 6, 9, 5, 7, 6, 5, 8, 6, 5, 5, 9, 8, 7, 4

2. On measurement of 200 metal bars the following distribution was formed.

Length (mm)	300	310	320	330	340	350	360	370	380	390
Frequency	2	10	21	37	65	41	15	5	3	1

(a) Which is the modal class?
(b) Find the value of the mode (i) by drawing
        (ii) by calculation.

3. The times taken for trains to travel between two stations were recorded for 280 journeys and are shown below.

Time taken (minutes)	64	65	66	67	68	69	70	71
Frequency	10	18	32	58	74	50	36	2

(a) Which is the modal class?
(b) Find the value of the mode   (i) by drawing
                (ii) by calculation.

6.2 THE MEDIAN

When a set of values are arranged in ascending or descending order of magnitude the middle value is the median. When the set contains an odd number of values the median is the middle value. When the set contains an even number of values the median is the average of the two middle values.

Example 6.4 Find the median of each of the following distributions:

(a) 4, 6, 7, 9, 10, 12, 16
(b) 3, 5, 7, 8, 11, 14, 15, 19
(c) 3, 19, 11, 5, 8, 7, 12, 5, 8, 15, 9

(a) the median is the value of the fourth variable, it is 9.

(b) There are eight variables, hence the median is the average of the fourth and fifth variables. Median = $\frac{8 + 11}{2} = 9.5$

(c) The variables must first be written in order of magnitude, 3, 5, 5, 7, 8, 8, 9, 11, 12, 15, 19. The median is the value of the sixth variable, it is 8.

When the distribution is given as a frequency table the median can be found.
1. By calculation from the frequency table,
2. From a histogram since the median divides the histogram into two equal parts,
3. From a cumulative frequency curve.

Example 6.5 The breaking loads of 90 steel rods were determined and the results recorded in the table below.

Breaking load (kN)	27.6	27.7	27.8	27.9	28.0	28.1	28.2	28.3
Frequency	2	4	14	21	23	18	7	1

Find the median from (a) the frequency table, (b) a histogram, (c) a cumulative frequency curve.

(a) There are 90 values and hence the median is the breaking load of the 45.5 rod. This will be in the 28.0 kN class. At the beginning of this class the breaking load will be 27.95 kN and at the end 28.05 kN. We assume that the 23 rods are equally spaced in this class. Since up to the beginning of this class there are $2 + 4 + 14 + 21 = 41$ rods we need to find the breaking load of the $45.5 - 41 = 4.5$ rod of the 23 in the class

41 rods	4.5 rods	18.5 rods	36 rods
27.95 kN	median	28.05 kN	

$$\text{Median} = 27.95 + \frac{4.5}{4.5 + 18.5} \times 0.1 = 27.95 + 0.0196$$

Median = 27.97 kN

(b) The histogram is drawn in fig. 6.3

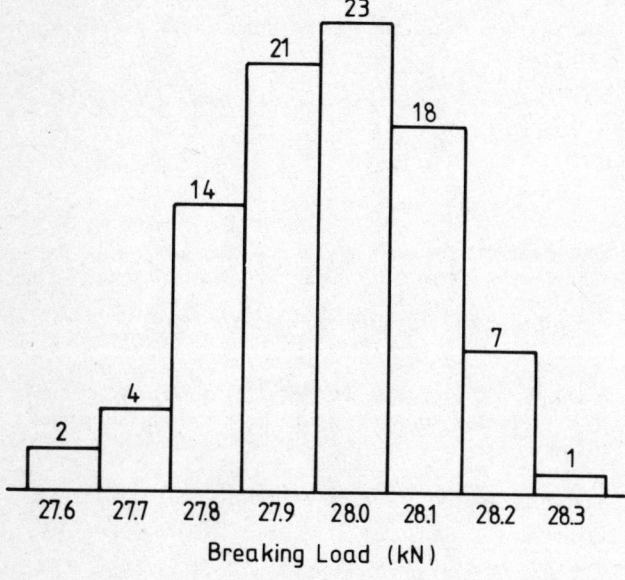

Figure 6.3

The median divides the histogram into two equal parts. The median will be in the class 27.95 to 28.05. If the median group is divided into x and $23 - x$ then

$$2 + 4 + 14 + 21 + x = 1 + 7 + 18 + (23 - x)$$

$$41 + x = 49 - x$$

$$2x = 8$$

$$x = 4$$

Area $x = 4$ square units and the area of the median group is 23 square units.

Therefore the width of $x = \dfrac{4}{23} \times$ class interval

$$= \frac{4}{23} \times 0.1 \text{ kN}$$

$$= 0.0174 \text{ kN}$$

$$\text{Median} = 27.95 + 0.02 = 27.97 \text{ kN}.$$

(c) A cumulative frequency curve is drawn by adding the frequencies together. The cumulative frequency curve gives the number of variables up to the end of the class. Hence the points for the cumulative frequency curve must be plotted at the end of the class interval.

Breaking load (kN)	27.6	27.7	27.8
Class interval	27.55–27.65	27.65–27.75	27.75–27.85
Frequency	2	4	14
Cumulative frequency	2	6	20
Percentage	2.2	6.7	22.2

27.9	28.0	28.1
27.85–27.95	27.95–28.05	28.05–28.15
21	23	18
41	64	82
45.6	71.1	91.1

28.2	28.3
28.15–28.25	28.25–28.35
7	1
89	90
98.9	100

The cumulative frequency curve is drawn in fig. 6.4.

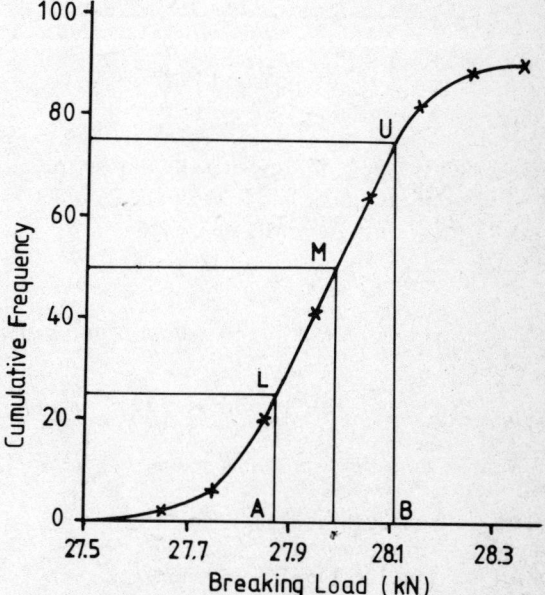

Figure 6.4

The median is found by drawing the horizontal line at a cumulative frequency of 45 and then drawing from the curve at the point P down to the horizontal axis. The median is 27.97 kN.

Similarly two lines are drawn at cumulative frequencies of 22.5 and 67.5 to points L and U. These lines divide the diagram into quarters. The part below the line at L is called the lower quartile and the part above the

line at U the upper quartile. The distance AB is called the interquartile range.
The interquartile range = 0.20 kN.
Sometimes the cumulative frequencies are converted to percentages and each unit on the vertical axis is then a percentile. The cumulative curve using percentiles is drawn in fig. 6.5.

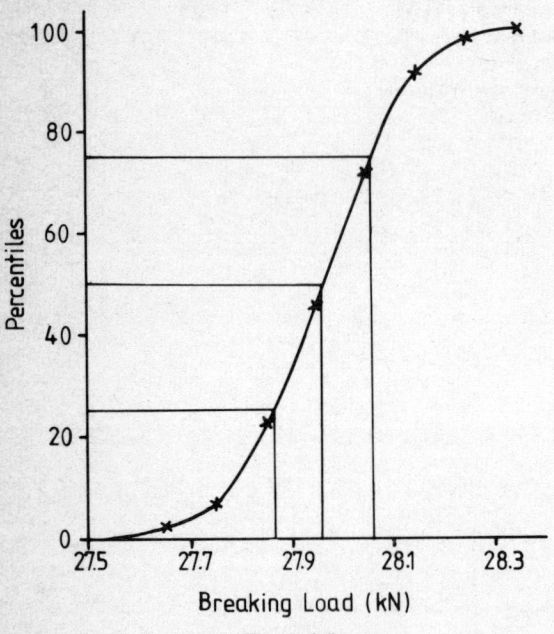

Breaking Load (kN)

Figure 6.5

The median is found by drawing the line at the 50th percentile and the upper and lower quartiles are drawn at the 75th and 25th percentiles respectively.

EXERCISE 6.2

1. Find the median of each of the following distributions:
(a)   1,   1,   3,   5,   7,   7,   8,   8,   9
(b)   2,   3,   3,   3,   4,   5,   5,   6,   6,   7
(c)   3,   7,   4,   4,   10,   11,   6,   4,   3,   7
(d)   11,   10,   4,   5,   9,   7,   6,   8,   5,   3

2. The times to complete a certain process are recorded in the frequency table below.

Time(s)	120–125	125–130	130–135	135–140
Frequency	3	8	15	34
Time(s)	140–145	145–150	150–155	
Frequency	23	5	2	

Find the median from (a) the frequency table, (b) a histogram.

3. The diameters of 120 rods were measured and the results recorded in the frequency table below.

Diameter (mm)	8.3	8.5	8.7	8.9	9.1	9.3	9.5
Frequency	4	24	36	30	16	6	4

Form a cumulative frequency table and from it draw a cumulative frequency curve. From the curve find (a) the median (b) the interquartile range.

6.3 THE MEAN

The mean of a distribution is obtained by dividing the sum of the variables by the number of variables.

$$\text{Mean} = \frac{\text{sum of variables}}{\text{number of variables}}$$

In statistics this is written $\bar{x} = \dfrac{\Sigma x}{n}$

where $\bar{x}$ is the mean value.
Σ is used as an abbreviation for 'the sum of all the terms like'
x represents the variables
n is the total frequency.

Example 6.6 Find the mean mass of the following castings
6.36,   6.47,   6.21,   6.38,   6.42,   6.34 and 6.43 kg
$\Sigma x = 6.36 + 6.47 + 6.21 + 6.38 + 6.42 + 6.34 + 6.43$
$\qquad = 44.61$
$n = 7$
$x = \dfrac{44.61}{7} = 6.37 \text{ kg}$

The mean mass is 6.37 kg.
When we want to find the mean of a frequency distribution then the formula becomes $x = \dfrac{\Sigma fx}{\Sigma f}$

Example 6.7 The thickness of 120 metal components was measured and recorded in the frequency table below.

Thickness (mm)	x	9.8	9.9	10.0	10.1	10.2	10.3	10.4
Frequency	f	4	14	17	26	28	21	10

Calculate the mean thickness.

$\Sigma fx = 4 \times 9.8 + 14 \times 9.9 + 17 \times 10.0 + 26 \times 10.1 +$
$\qquad\qquad 28 \times 10.2 + 21 \times 10.3 + 10 \times 10.4$
$\Sigma fx = 1216.3$
$\Sigma f = 120$
$\bar{x} = \dfrac{1216.3}{120} = 10.14 \text{ mm}$

The mean = 10.14 mm

This calculation can be done directly on a calculator by accumulating the products 4×9.8, 14×9.9 etc. in the memory and dividing the final sum by 120.

Example 6.8 The lengths of a number of components were measured and recorded in the table below.

Length (mm)	x	221–224	224–227	227–230	230–233
Frequency	f	7	14	37	47

233–236	236–239
23	11

Calculate the mean length.

We assume that each member in a class has the length of the middle of the class. The middle of the first class is 222.5 mm and of the second class 225.5 mm etc.

$\Sigma fx = 7 \times 222.5 + 14 \times 225.5 + 37 \times 228.5 +$
$\qquad 47 \times 231.5 + 23 \times 234.5 + 11 \times 237.5$
$\Sigma fx = 32\,055.5$
$\Sigma f = 7 + 14 + 37 + 47 + 23 + 11 = 139$
$\bar{x} = \dfrac{32\,055.5}{139} = 230.6 \text{ mm}$

The mean length = 230.6 mm.
The above calculation can be done by using an assumed mean and a table. This makes the arithmetic easier.
Take as the assumed mean, a, the value of x which has the largest frequency.
$a = 231.5$

x	f	$x - a$	$f(x - a)$
222.5	7	-9	-63
225.5	14	-6	-84
228.5	37	-3	-111
231.5	47	0	0
234.5	23	3	69
237.5	11	6	66
			$f(x-a) = -123$

$\bar{x} = a + \dfrac{\Sigma f(x - a)}{n} = 231.5 + \dfrac{-123}{139}$

$\bar{x} = 231.5 - 0.88$
$\bar{x} = 230.6 \text{ mm}$

Example 6.9 The marks awarded to 150 children in an examination were as follows:

Marks	1–10	11–20	21–30	31–40	41–50	51–60
Frequency	3	10	14	24	26	26

	61–70	71–80	81–90	91–100
	25	12	6	4

Calculate the mean mark
Using an assumed mean method we take $a = 45.5$

Marks	Mid value x	Frequency f	$x - a$	$f(x - a)$
1–10	5.5	3	-40	-120
11–20	15.5	10	-30	-300
21–30	25.5	14	-20	-280
31–40	35.5	24	-10	-240
41–50	45.5	26	0	0
51–60	55.5	26	10	260
61–70	65.5	25	20	500
71–80	75.5	12	30	360
81–90	85.5	6	40	240
91–100	95.5	4	50	200
		$n = 150$		$\Sigma f(x - a) = 620$

$\bar{x} = a + \dfrac{\Sigma f(x - a)}{n} = 45.5 + \dfrac{620}{150}$

$\bar{x} = 45.5 + 4.13 = 49.63$
The mean mark was 49.63
The mean can also be calculated directly on a calculator by accumulating the products of x and f in the memory and dividing by the total frequency.

CM $\underset{\smile}{x} \times f = $ M+ RM $\div n =$

EXERCISE 6.3

1. The following are the times taken to complete a certain process. 2.3, 2.7, 2.5, 3.1, 2.4, 2.7, 2.8, 2.4, 2.9, 3.0, 2.8 and 2.7 hours. Calculate the mean time.
2. The lengths of a batch of metal bars in metres were 1.61, 1.58, 1.63, 1.63, 1.60, 1.59, 1.63, 1.62, 1.61 and 1.59. Find the mean length.

3. The diameters of 120 rods were measured and the results recorded in the frequency table below.

Diameter (mm)	8.3	8.5	8.7	8.9	9.1	9.3	9.5
Frequency	4	24	36	30	16	6	4

Calculate the mean diameter.

4. The times to complete a certain process are recorded in the frequency table below.

Time (s)	120–125	125–130	130–135	135–140
Frequency	3	8	15	34

	140–145	145–150	150–155
	23	5	2

Calculate the mean time.

6.4 GENERAL EXAMPLES

EXERCISE 6.4

1. The lives of 100 electric light bulbs were recorded in the frequency table below

Life (hours)	925	950	975	1000	1025	1050	1075	1100
Frequency	2	7	14	31	37	6	2	1

Draw a histogram to represent this data and find the mode, median and mean of the distribution.

2. The fusing current for 170 wires was recorded and grouped in the frequency table below.

Fusing current (A)	4.15–4.20	4.20–4.25	4.25–4.30
Frequency	12	23	35

	4.30–4.35	4.35–4.40	4.40–4.45	4.45–4.50
	41	32	19	8

Find the mode, median and mean. Draw a cumulative frequency curve and from it estimate the median and the interquartile range.

3. Screws were packed in boxes, the nominal number being 150. 100 boxes were checked and the numbers recorded in the table below.

Number of screws	147	148	149	150	151	152	153	154	155
Frequency	3	6	9	14	15	20	18	10	5

Draw a histogram to represent this data and find the mode, median and mean of the distribution.

SUMMARY

1. The mode of a set of data is the value which occurs most frequently. It can be estimated from a histogram.

2. When the variables of a set of data are arranged in ascending order or descending order the value of the middle variable is called the median. When there are an even number of variables the median is the average of the middle two values. The median can be found from a cumulative frequency curve.

3. The mean of a set of data is the sum of the variables in the data divided by the number of variables.

$$\bar{x} = \frac{\Sigma x}{n} \quad \text{or} \quad \bar{x} = \frac{\Sigma fx}{\Sigma f}$$

Where x represents the variables, n the number of variables, f the frequency and Σ is an abbreviation for 'the sum of all the terms like'.

SELF ASSESSMENT PAPER No 6

Answer all the questions
Time allowed: Section A 10 minutes (2 marks each question)
Section B 50 minutes (10 marks each question)
Marks gained: 19+ pass with credit, 15—19 pass, less than 15 fail, repeat chapter 6.

Section A

1. Find the mode of 2, 7, 9, 8, 7, 8, 6, 5, 7, 2, 4

2. Find the median of 3, 3, 4, 4, 6, 7, 7, 8

3. Find the median of 5, 3, 8, 6, 7, 9, 8, 9, 10, 8, 3

4. Find the mean of 9, 8, 8, 7, 6, 6, 4, 4

Section B

1. The diameters of 120 rivet heads were measured and the results are recorded in the frequency table below.

Diameter (mm)	8.4–8.6	8.6–8.8	8.8–9.0	9.0–9.2	9.2–9.4
Frequency	3	9	19	35	29

9.4–9.6	9.6–9.8
17	8

Draw a histogram and find the mode of the distribution.

2. The speeds in kilometres per hour of 800 vehicles passing along a road were measured and the results grouped in the following table:

Mid interval speed (km/h)	25	30	35	40	45	50	55	60
Frequency		6	24	75	120	181	254	120	20

Illustrate these results by drawing a cumulative frequency curve. From the curve estimate the median speed and the interquartile range of the distribution.

3. 120 cubes of concrete were found to have crushing loads measured in Newtons as follows.

Load (N)	6500	7000	7500	8000	8500	9000	9500
Frequency	3	10	22	25	28	19	13

Calculate the mean crushing load.

ANSWERS

Exercise 6.1   1. (a) 11, (b) 5 and 8, (c) 5,   2. (a) 340, (b)   (i) 340 (ii) 340.4,   3. (a) 68, (b)   (i) 67.9, (ii) 67.9.

Exercise 6.2   1.(a) 7, (b) 4.5, (c) 5, (d) 6.5, 2.(a) 137.8 s, (b) 137.8 s   3. (a) 8.78 mm, (b) 0.35 mm.

Exercise 6.3   1. 2.69 s   2. 1.61 m   3. 8.81 mm, 4. 137.4 s.

Exercise 6.4   1. 1016.6, 1016.6, 1006,3,   2. 4.32, 4.32, 4.32, 4.32, 0.114,   3. 152, 152, 151

SELF ASSESSMENT PAPER No 6

	Marks
Section A	
1. 2, 2, 4, 5, 6, 7, 7, 7, 8, 8, 9	1
mode = 7	1
2. Median = $\frac{1}{2}(4 + 6) = 5$	2
3. 3, 3, 5, 6, 7, 8, 8, 8, 9, 9, 10	1
median = 8	1
4. Mean = $\dfrac{9 + 8 + 8 + 7 + 6 + 6 + 4 + 4}{8}$	1
$= 6.5$	1

	Marks
Section B	
1. Histogram	5
Mode $= \dfrac{16}{16 + 6} \times 0.2 + 9$	3
$= 9.15$	2

	Marks

2. End of interval   27.5   32.5   37.5   42.5   47.5

   Cumulative

   frequency   6   30   105   225   406

                     52.5 57.5 62.5

                     660   780   800   2

Cumulative frequency curve plotted from
above table   4

Median = 47.2 km/h   2

Interquartile range = 51.5 − 41.7   1

                 = 9.8 km/h   1

	Marks

3. Mean $= \dfrac{3 \times 6500 + 10 \times 7000 \text{ etc.}}{3 + 10 + 22 \text{ etc.}}$ 5

        $= \dfrac{987\,000}{120}$ 4

        $= 8225$ N   1

7 The Standard Deviation

7.1 UNGROUPED DATA

In chapter 6 we looked at the mode, median and the mean. These are all measures of central tendency. These measures of the centre of a distribution do not tell us anything about the way the variables are spread about the mean. For example the numbers 9, 10, 10, 11, 12, 12, 13 have a mean of 11, also the numbers 3, 7, 9, 12, 14, 15, 17 have a mean of 11. The first set of numbers are all near to the mean, the second set are more widely spread. The range of a distribution is the difference between the highest and lowest variables. In the first distribution the range is $13 - 9 = 4$, in the second it is $17 - 3 = 14$. Hence the range can tell us how the variables are spread about the mean, but since it only involves the highest and lowest variables it is not very reliable.

The best measure of spread is the standard deviation. It takes into account the deviation of each variable and hence is more reliable. When the variables are a long way from the mean the standard deviation is large. The standard deviation is the square root of the average of the squares of the differences between mean and the variables.

If the variables are $x_1, x_2, x_3, \ldots x_n$ then the standard deviation is represented by σ where

$$\sigma = \sqrt{\frac{(x_1 - \bar{x})^2 + (x_2 - \bar{x})^2 + (x_3 - \bar{x})^2 + \ldots (x_n - \bar{x})^2}{n}}$$

σ is sigma, one of the lower case letters of the Greek alphabet.

Example 7.1 Find the mean and the standard deviation of the following masses, all in kg, 12.3, 12.5, 12.8, 12.7, 12.7, 12.9, 12.9, 13.0, 12.6
The mean,

$$\bar{x} = \frac{12.3 + 12.5 + 12.8 + 12.7 + 12.7 + 12.9 + 12.9 + 13.0 + 12.6}{9}$$

$$\bar{x} = \frac{114.4}{9} = 12.7$$

The standard deviation σ is given by

$$\sigma^2 = \frac{(12.3 - 12.7)^2 + (12.5 - 12.7)^2 + \ldots (12.6 - 12.7)^2}{9}$$

$$\sigma = \sqrt{\frac{0.39}{9}} = \sqrt{0.043\,33} = 0.21$$

The mean mass is 12.7 kg and the standard deviation is 0.21 kg.
Note: the standard deviation has the same units as the variables. The value of σ^2 is called the variance.

Example 7.2 Mr Jones travelled the same way to work each day. The times taken on 9 consecutive journeys were 17.8, 17.4, 18.2, 18.8, 16.9, 17.6, 19.5, 17.5 and 18.3 minutes. Calculate the mean time for the journey and the standard deviation. Which of the times were more than one standard deviation from the mean?

The mean $\bar{x} =$

$$\frac{17.8 + 17.4 + 18.2 + 18.8 + 16.9 + 17.6 + 19.5 + 17.5 + 18.3}{9}$$

$$\bar{x} = 18.0$$

The standard deviation $=$

$$\sqrt{\frac{(17.8 - 18.0)^2 + (17.4 - 18.0)^2 + \ldots}{9}}$$

$$= \sqrt{\frac{0.04 + 0.36 + 0.04 + 0.64 + 1.21 + 0.16 + 2.25 + 0.25 + 0.}{9}}$$

$$= \sqrt{\frac{5.04}{9}} = \sqrt{0.56} = 0.748$$

The mean time is 18.0 min and the standard deviation is 0.75 min. The mean plus one standard deviation $= \bar{x} + \sigma = 18.0 + 0.75 = 18.75$. The mean minus one standard deviation $= \bar{x} - \sigma = 18.0 - 0.75 = 17.25$
The times outside $\bar{x} + \sigma$ and $\bar{x} - \sigma$ are those greater than 18.75 min or less than 17.25 min.
Hence the times are 18.8 min, 19.5 min, 16.9 min

EXERCISE 7.1

1. Calculate the mean and standard deviation for the following sets of data (a) the masses of 7 castings are 7.2, 7.3, 6.9, 7.2, 7.5, 7.3 and 7.4 kg.
(b) The times to complete a certain production process were 56, 53, 55, 58, 54, 59, 55 and 53 s.
(c) The lengths of 10 components measured in mm were 163, 162, 167, 161, 164, 165, 163, 168, 162 and 168.

2. The times to evacuate a building after the sounding of the fire alarm were recorded on 12 different occasions, the times in seconds being 47, 52, 57, 43, 61, 59, 43, 51, 48, 53, 54 and 49. Calculate the mean time and the standard deviation from the mean.
Which times lie between ± 1 standard deviation from the mean?

7.2 GROUPED DATA

When finding the standard deviation for grouped data each $(x - \bar{x})^2$ term must be multiplied by its respective frequency. Hence the standard deviation σ is given by

$$\sigma = \sqrt{\frac{\Sigma f(x - \bar{x})^2}{\Sigma f}}$$

When using a calculator this formula can be written

$$\sigma = \sqrt{\frac{\Sigma fx^2}{\Sigma f} - (\bar{x})^2}$$

but much care must be taken since this calculation involves the difference of two large numbers which can be almost equal.
If a calculator is not used it is easier to use an assumed mean to calculate the mean and the standard deviation. If the assumed mean is a then

$$\bar{x} = \frac{\Sigma f(x - a)}{\Sigma f} + a, \qquad \sigma = \sqrt{\frac{\Sigma f(x - a)^2}{\Sigma f} - (\bar{x} - a)^2}$$

Example 7.2 A sample of 90 manufactured components with a nominal length of 160 mm as measured and the lengths recorded in the table below.

Length x (mm) 154 156 158 160 162 164 166 168
Frequency f 4   12   20   26   16   7   4   1

Calculate the mean length and the standard deviation.
Method 1. Using an assumed mean $a = 160$

x	f	$x - a$	$f(x - a)$	$f(x - a)^2$
154	4	− 6	− 24	144
156	12	− 4	− 48	192
158	20	− 2	− 40	80
160	26	0	0	0
162	16	2	32	64
164	7	4	28	112
166	4	6	24	144
168	1	8	8	64
	90		− 20	800

$\Sigma f = 90$, $\Sigma f(x - a) = -20$, $\Sigma f(x - a)^2 = 800$

$$\bar{x} = \frac{\Sigma f(x - a)}{\Sigma f} + a = \frac{-20}{90} + 160 = 159.8 \text{ mm}$$

$$\sigma = \sqrt{\frac{\Sigma f(x - a)^2}{\Sigma f} - (\bar{x} - a)^2} = \sqrt{\frac{800}{90} - \left(\frac{-20}{90}\right)^2}$$

$$\sigma = \sqrt{8.889 - 0.049} = \sqrt{8.840}$$

$$\sigma = 2.97 \text{ mm}$$

Method 2. Using a calculator

$$x = \frac{\Sigma fx}{\Sigma f} = \frac{14\,380}{90} = 159.778 \text{ mm}$$

$$\sigma = \sqrt{\frac{\Sigma fx^2}{\Sigma f} - \bar{x}^2} = \sqrt{\frac{2\,298\,400}{90} - (159.778)^2}$$

$$\sigma = \sqrt{25\,537.772 - 25\,529.009} = \sqrt{8.768}$$
$$= 2.96 \text{ mm}$$

The mean length was 159.8 mm and the standard deviation was 2.97 mm.

As was stated previously great care must be taken when using the calculator method for the standard deviation. For example in the above calculation if the mean is rounded to 159.8 for the standard deviation calculation a value of $\sigma = 1.32$ will be obtained. In the above example the centre of each group is given. If the groups are given as ranges then the middle of the range is taken when calculating the mean or standard deviation. For example if the groups are 10−20, 20−30, 30−40, etc. then the centres are 15, 25, 35, etc.

EXERCISE 7.2

1. A survey was made of the earnings of 1000 people, the results are recorded in the frequency table below

Earnings (£)   90 95 100 105 110 115 120 125
Frequency   40 60 100 140 220 240 140   60

Calculate the mean and the standard deviation.

2. The lengths of 200 components are measured and the lengths recorded in the distribution table below

Length (mm) 17.1 17.2 17.3 17.4 17.5 17.6
Frequency   5   12   26   32   45   35

17.7 17.8 17.9
 22   19   4

Calculate the mean length and the standard deviation.

3. The lives of 150 electric light bulbs were measured and the results recorded in a frequency table

Life (hours)   600−700 700−800 800−900 900−1000
Frequency   11   23   45   37

1000−1100 1100−1210
   28   6

Calculate the mean life and the standard deviation of the distribution.

4. The thickness of a sample of 100 steel plates was measured.

Thickness (mm) 7.6−7.7 7.7−7.8 7.8−7.9 7.9−8.0 8.0−8.1
Frequency   3   8   12   25   30

8.1−8.2 8.2−8.3
  14   8

Calculate the mean thickness and the standard deviation from the mean.

7.3 GENERAL EXAMPLES

Example 7.4 Two machines are producing pins to the same specification. The nominal diameter of the pins is 20 mm. The diameters of 60 pins from each machine were measured and the results recorded in the frequency table below:

Machine A

Diameter (mm)	19.97	19.98	19.99	20.00	20.01	20.02	20.03
Frequency	4	8	12	14	11	7	4

Machine B

Diameter (mm)	19.97	19.98	19.99	20.00	20.01	20.02	20.03
Frequency	1	7	8	25	10	6	3

Calculate the mean and the standard deviation for each distribution and compare the results.

Machine A take $a = 20$

x	f	$x - a$	$f(x - a)$	$f(x - a)^2$
19.97	4	− 0.03	− 0.12	0.0036
19.98	8	− 0.02	− 0.16	0.0032
19.99	12	− 0.01	− 0.12	0.0012
20.00	14	0	0	0
20.01	11	0.01	0.11	0.0011
20.02	7	0.02	0.14	0.0028
20.03	4	0.03	0.12	0.0036
	60		− 0.03	0.0155

$\Sigma f = 60, \quad \Sigma f(x - a) = - 0.03, \quad \Sigma f(x - a)^2 = 0.0155$

$$\bar{x} = \frac{\Sigma f(x - a)}{\Sigma f} + a = \frac{-0.03}{60} + 20.00 = 20.00$$

$$\sigma = \sqrt{\frac{\Sigma f(x - a)^2}{\Sigma f} - (\bar{x} - a)^2}$$

$$= \sqrt{\frac{0.0155}{60} - (20.00 - 20.00)^2}$$

$$\sigma = \sqrt{0.000\,258\,3} = 0.016$$

Machine B take $a = 20$

x	f	$x - a$	$f(x - a)$	$f(x - a)^2$
19.97	1	− 0.03	− 0.03	0.0009
19.98	7	− 0.02	− 0.14	0.0028
19.99	8	− 0.01	− 0.08	0.0008
20.00	25	0	0	0
20.01	10	0.01	0.10	0.0010
20.02	6	0.02	0.12	0.0024
20.03	3	0.03	0.09	0.0027
	60		0.06	0.0106

$\Sigma f = 60, \quad \Sigma f(x - a) = 0.06, \quad \Sigma f(x - a)^2 = 0.0106$

$$\bar{x} = \frac{\Sigma f(x - a)}{\Sigma f} + a = \frac{0.06}{60} + 20.00 = 20.00$$

$$\sigma = \sqrt{\frac{\Sigma f(x - a)^2}{\Sigma f} - (\bar{x} - a)^2}$$

$$= \sqrt{\frac{0.0106}{60} - (20.00 - 20.00)^2}$$

$$\sigma = \sqrt{0.00\,176\,6} = 0.013$$

Machine A is producing pins with a mean diameter of 20.00 mm and a standard deviation of 0.016 mm. Machine B is producing pins with a mean diameter of 20.00 mm and a standard deviation of 0.013 mm. Hence although both machines are producing pins with a mean of 20.00 mm it appears that machine B is working more accurately. This can be seen from the diagram, in fig. 7.1, which shows the distributions.

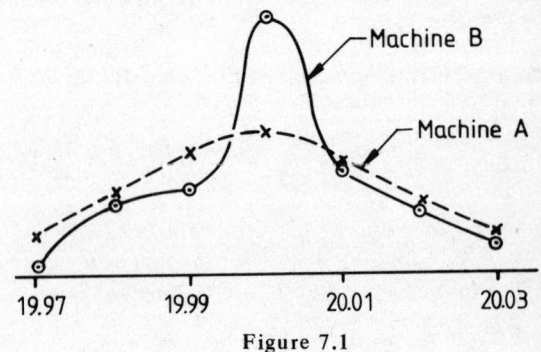

Figure 7.1

EXERCISE 7.3

1. The capacities of 11 barrels were recorded in litres as follows: 50.4, 50.2, 50.2, 49.8, 50.1, 50.2, 50.4, 49.7, 50.5, 50.2, 49.6. Calculate the mean capacity and the standard deviation. If all barrels which were less than one standard deviation below the mean or more than two standard deviations above the mean were rejected, which of these barrels would be rejected?

2. 100 specimens of wire were tested for their fusing current, the results are given in the following frequency table

Fusing current (A)	29.4	29.6	29.8	30.0	30.2	30.4	30.6
Frequency	5	14	22	28	19	9	3

Calculate the mean fusing current and the standard deviation.

3. A sample of 10 items was taken from the days production made on machine A. The lengths of the items in the sample were 7.73, 7.84, 7.76, 7.79, 7.74, 7.81, 7.74, 7.77, 7.82, 7.70 mm. Calculate the mean length and the standard deviation from the mean for this sample.
Another machine B is producing the same item. A sample of 10 from this machine was found to have a mean of 7.78 and a standard deviation of 0.120. From the means and standard deviations of these two samples which machine appears to be (a) working most accurately, (b) producing items of greater length?

4. Two machines are packing sugar in 1 kg cartons. 100 samples are taken from each machine, the results are shown below.

Machine A

Mass of sugar (kg)	0.97	0.98	0.99	1.00	1.01	1.02	1.03	1.04
Frequency	2	6	13	16	23	17	12	11

Machine B

Mass of sugar (kg)	0.96	0.97	0.98	0.99	1.00	1.01	1.02	1.03
Frequency	1	1	7	19	29	26	15	2

Find the means and standard deviations for the two distributions and compare the results pointing out any differences between the outputs from the two machines.

SUMMARY

1. The standard deviation of a set of variables x_1, x_2, $x_3, x_4, \ldots x_n$ is

$$\sigma = \sqrt{\frac{(x_1 - \bar{x})^2 + (x_2 - \bar{x})^2 + (x_3 - \bar{x})^2 \ldots + (x_n - \bar{x})^2}{n}}$$

$\bar{x}$ is the mean of $x_1, x_2, \ldots x_n$ and n is the number of variables.

2. When the variables are grouped into a frequency distribution then

$$\sigma = \sqrt{\frac{\Sigma f(x - \bar{x})^2}{\Sigma f}}$$

$$= \sqrt{\frac{\Sigma f(x - a)^2}{\Sigma f} - (\bar{x} - a)^2} = \sqrt{\frac{\Sigma f x^2}{\Sigma f} - \bar{x}^2}$$

(without a calculator)　　　(with a calculator)

where Σf is the sum of the variables and a is an assumed mean.

3. The standard deviation is a measure of the spread of the distribution about the mean. When the standard deviation is high the variables are spread widely about the mean. When the standard deviation is low the variables are close to the mean.

SELF ASSESSMENT PAPER No 7

Instructions: Answer all questions
Time allowed: One hour (20 marks each question)
Marks gained: 30+ pass with credit, 24–30 pass, less than 24 fail, repeat chapter 7.

1. A machine is filling 5 litre cans with oil. 10 cans are selected and the volume of oil in the can measured. The results were 5.03, 5.01, 5.01, 5.04, 5.01, 4.97, 4.96, 5.00, 5.03 and 5.01 litres.
Calculate the mean volume of oil and the standard deviation. If cans with less than one standard deviation below the mean were rejected, which, if any, of the cans were rejected?

2. A sample of 100 resistors were taken and their resistance measured. The values of the resistances are recorded in the frequency table below.

Resistance (Ω)	146	147	148	149	150	151	152	153	154
Frequency	1	4	12	20	32	18	9	3	1

Calculate the mean and standard deviation.

3. The marks awarded to 150 students in an examination were as follows:

Marks	1–10	11–20	21–30	31–40	41–50	51–60
Frequency	3	10	14	24	26	27

61–70	71–80	81–90	91–100
24	14	4	4

Calculate the mean mark awarded and the standard deviation.

ANSWERS

Exercise 7.1 1.(a) 7.26 kg, 0.18 kg, (b) 55.4 s, 2.05 s, (c) 164.3 mm, 2.45 mm, 2. 51.4 s, 5.5 s; 47, 52, 51, 48, 53, 54 and 49 s.

Exercise 7.2 1. £110.10, £8.75, 2. 17.5 mm, 0.185 mm, 3. 894, 127, 4. 8.0 mm, 0.144 mm.

Exercise 7.3 1. 50.1 litres, 0.28 litres, 49.8, 49.7, 49.6 litres, 2. 29.96 A, 0.284 A, 3. 7.77 mm, 0.042 mm, (a) Machine A, (b) Machine B, 4. Machine A, 1.0106 kg, 0.0178 kg, Machine B 1.0022 kg, 0.0131 kg; Machine B appears to be working more accurately.

SELF ASSESSMENT PAPER No 7

	Marks
1. $\bar{x} = \dfrac{5.03 + 5.01 + \ldots}{10}$	3
$\bar{x} = 5.01$ litres	2
$\sigma = \sqrt{\dfrac{(5.03 - 5.01)^2 + (5.01 - 5.01)^2 + \ldots}{10}}$	5
$\sigma = 0.024$	5
$\bar{x} - \sigma = 4.986$ litres	3
The 4.97 and 4.96 litre cans are rejected	2
2. Either by calculator $\bar{x} = 149.9 \, \Omega$	6
$\sigma = 1.47 \, \Omega$	14

or

x	f	$x - a$	$f(x - a)$	$f(x - a)^2$
146	1	−4	−4	16
147	4	−3	−12	36
148	12	−2	−24	48
149	20	−1	−20	20
150	32	0	0	0

Marks

2. (*continued*)

x	f	$x-a$	$f(x-a)$	$f(x-a)^2$
151	18	1	18	18
152	9	2	18	36
153	3	3	9	27
154	1	4	4	16
	100		-11	217

$a = 150$, $\Sigma f = 100$, $\Sigma f(x-a) = -11$,

$\Sigma f(x-a)^2 = 217$ — 6

$\bar{x} = 150 + \dfrac{-11}{100} = 149.9\,\Omega$ — 4

$\sigma = \sqrt{\dfrac{217}{100} - \left(\dfrac{-11}{100}\right)^2}$ — 3

$\sigma = 1.47\,\Omega$ — 3

3. Either by calculator $\bar{x} = 49.43$ — 6

$\sigma = 20.13$ — 14

or

x	f	$x-a$	$f(x-a)$	$f(x-a)^2$
5.5	3	-50	-150	7500
15.5	10	-40	-400	16000
25.5	14	-30	-420	12600
35.5	24	-20	-480	9600
45.5	26	-10	-260	2600
55.5	27	0	0	0
65.5	24	10	240	2400
75.5	14	20	280	5600
85.5	4	30	120	3600
95.5	4	40	160	6400
	150		-910	66300

6

$a = 55.5$, $\Sigma f = 150$, $\Sigma f(x-a) = -910$,

$\Sigma f(x-a)^2 = 66\,300$ — 6

$\bar{x} = 55.5 - \dfrac{910}{150} = 49.43$ — 3

$\sigma = \sqrt{\dfrac{66\,300}{150} - \left(\dfrac{910}{150}\right)^2}$ — 2

$\sigma = 20.13$ — 3

EXAMINATION 1

Instructions: Answer all questions in Section A and 4 questions from Section B

Time allowed: Section A 30 minutes (30 marks)

Section B 90 minutes (20 marks each question)

Marks gained: 55+ pass with credit, 44–55 pass, less than 44 fail, repeat chapters 1–7.

Section A

1. Copy the table below and fill in the cumulative frequency values

Length (mm)	160	170	180	190	200
Frequency	7	29	38	25	4
Cumulative frequency					

2. (a) Give two examples of discrete variables.
(b) Give two examples of continuous variables.

3. (a) Find the standard deviation of 3.5, 3.7, 3.8, 4.0, 4.1, 4.2, 4.3 and 4.4.
(b) Name one other measure of dispersion besides standard deviation.

4. Given $y = ax^2 + b$.
(a) Transpose the formula to make x the subject.
(b) Calculate y when $a = -2.13$, $x = 1.84$ and $b = 12.5$.

5. Which of the following triangles can be solved by using the sine rule. (a) Given a, b, C find c, (b) Given a, b, A find B, (c) given a, b, c find A, (d) Given a, B, C find b.

Section B

1. A sample of eighty rods from the initial output of a production line was measured to the nearest millimetre and gave the following results.

Length (mm)	180–181	182–183	184–185	186–187
Number of rods	4	12	27	18

	188–189	190–191	192–193
	10	7	2

(i) State the greatest and least possible lengths of rod in the sample.
(ii) Form a cumulative frequency table and from it construct a cumulative frequency curve.
(iii) What is the median value?
(iv) If a rod is rejected if it is less than 182.5 mm or more than 192.5 mm, what percentage are rejected?

2. The information below relates to the road life of tyres. Calculate the mean life and the standard deviation life.

Distance covered (km)	Number of tyres
15 000–17 500	2
17 500–20 000	4
20 000–22 500	40
22 500–25 000	130
25 000–27 500	190
27 500–30 000	110
30 000–32 500	20
32 500–35 000	4
total	500

Marks

3. Given the formula $V = \dfrac{\pi h}{3}(r^2 + R^2 + rR)$.

(a) Transpose the formula to make h the subject.
(b) Write out a sequence of steps you would use to calculate V.
(c) Calculate the value of V when $h = 62\,\text{mm}$, $r = 22\,\text{mm}$ and $R = 37\,\text{mm}$.

4. Draw the graph of $y = 4x^3 - 8x^2 - 15x + 9$ for values of x between -2 and $+3$. From the graph write down the solutions to the equation $4x^3 - 8x^2 - 15x + 9 = 0$.

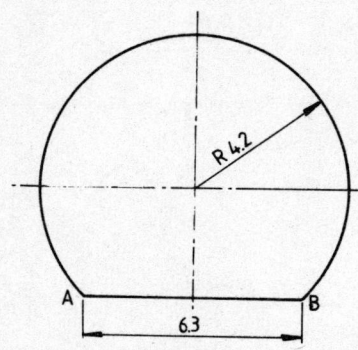

Figure 1

5. Figure 1 shows the section of a tunnel which consists of a major segment of a circle. The radius of the circle is 4.2 m and the width of the base is 6.3 m. Calculate (a) the maximum height of the tunnel, (b) the cross-sectional area of the tunnel.

6. The resistance R ohms of a resistor is measured at various temperatures $t°\text{C}$. The results are shown below.

R (ohms)	168.8	172.4	176.7	179.1	183.5
t (°C)	42	48	55	59	66

Draw a graph to show that these values obey a law of the form $R = at + b$ and hence determine values for a and b.

ANSWERS EXAMINATION 1

Marks

Section A

1. Cumulative frequency 7 36 74 99 103   3

2. (a) Number of students, number of components etc.   2
(b) Length of component, height of students etc.   2

3. (a) $\bar{x} = \dfrac{3.5 + 3.7 + 3.8 \ldots}{8} = 4$ 3

$\sigma = \sqrt{\dfrac{(4 - 3.5)^2 + (4 - 3.7)^2 \ldots}{8}}$ 3

$\sigma = 0.292$ 2
(b) Range   2

	Marks
4. (a) $ax^2 = y - b$	2

$x^2 = \dfrac{y - b}{a}$ 2

$x = \sqrt{\dfrac{y - b}{a}}$ 2

(b) $y = -2.13 \times (1.84)^2 + 12.5$ 3
$y = 5.289$ 2

5. (b), (d)   2

Section B
1. (i) Least 179.59 mm; greatest 193.49 mm   4
(ii) cumulative frequencies 4, 16, 43, 61, 71, 78, 80   3
Cumulative frequency curve   8
(iii) Median value = 185.4 mm   2
(iv) Rejected 9 below 182.5 mm, 1 above 192.5 mm   2
= 12.5%   1

2.

Distance x	Frequency f	$x - a$	$f(x - a)$ $(\times 10^3)$	$f(x - a)^2$ $(\times 10^6)$
16 250	2	$-10\,000$	-20	200
18 750	4	$-7\,500$	-30	225
21 250	40	$-5\,000$	-200	1000
23 750	130	$-2\,500$	-325	812.5
26 250	190	0	0	0
28 750	110	2 500	275	687.5
31 250	20	5 000	100	500
33 750	4	7 500	30	225
	500		-175	3650

8

$a = 26\,250$, $\Sigma f(x - a) = -175 \times 10^3$
$\Sigma f(x - a)^2 = 3650 \times 10^6$ 3

$\bar{x} = 26\,250 + \dfrac{-175\,000}{500} = 25\,900\,\text{km}$ 3

$\sigma = \sqrt{\dfrac{3650 \times 10^6}{500} - \left(\dfrac{-175\,000}{500}\right)^2}$ 3

$\sigma = 2679\,\text{km}$ 3

3. (a) $3V = \pi h(r^2 + R^2 + rR)$ 4

$\dfrac{3V}{\pi(r^2 + R^2 + rR)} = h$ 4

(b) CM 22 $\times$ = M+ 37 $\times$ = M+ 22 $\times$ 37 = M+ RM $\div$ 3 $\times$ π $\times$ 62 =   6

(c) $V = \dfrac{\pi \times 62}{3}(22^2 + 37^2 + 22 \times 37)$ 4

$V = 173\,200\,\text{mm}^3$ 2

4. x	-2	-1	0	1	2	3
y	-25	12	9	-10	-21	0

6

	Marks
Graph of $y = 4x^3 - 8x^2 - 15x + 9$	8
The solutions are where the curve cuts the x-axis	
$x = -1.5, 0.5, 3$	6

5. (a) Distance of base from centre of circle

$$= \sqrt{4.2^2 - 3.15^2}$$ 2

$$= 2.778$$ 2

Maximum height $= 4.2 + 2.778 = 7.0\,\text{m}$ 2

(b) Angle at centre of circle subtended by base $= \theta$

$$\sin\frac{\theta}{2} = \frac{3.15}{4.2}$$ 4

$\theta = 97.18°$ (1.696 rad)   2

	Marks
Cross-sectional area = area of sector OACB	
+ area of $\triangle$OAB	1
$= \frac{1}{2} \times 4.2^2 \times (2\pi - 1.696) + \frac{1}{2} \times 6.3 \times 2.778$	4
$= 40.459 + 8.751 = 49.210\,\text{m}^2$	3

6. Graph   6

Two points from graph $t = 40, R = 167.5$ 2

$t = 70, R = 185.8$ 2

$185.8 = 70a + b$ (1)   2

$167.5 = 40a + b$ (2)   2

Subtract (2) from (1) $18.3 = 30a,\ a = 0.61$ 2

From (1) $185.8 = 70 \times 0.61 + b$ 2

$b = 143$ 2

8 Surface Areas and Volumes of Regular Figures

8.1 PYRAMIDS, CONES AND SPHERES

A pyramid has a regular base and the apex is vertically above the centre of the base. The most common pyramid has a square base. A cone is a special case of a pyramid when the base is a circle.

For any pyramid or cone the surface area of the sloping sides is equal to half the perimeter of the base times the slant height. The volume is one third of the base area times the perpendicular height.

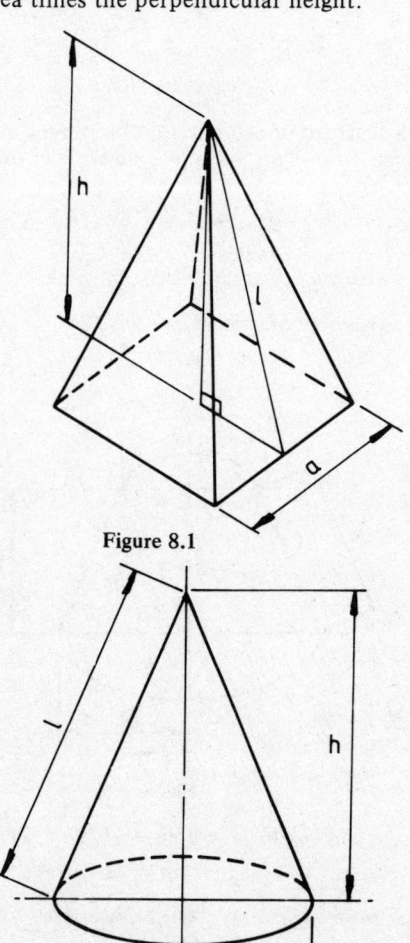

Figure 8.1

Figure 8.2

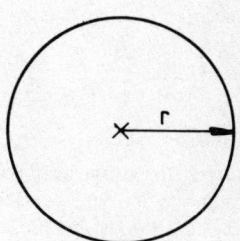

Figure 8.3

For a square pyramid, see fig. 8.1

$$\text{Area of sides} = \tfrac{1}{2} \times 4a \times l = 2al$$
$$\text{Volume} = \tfrac{1}{3} \times a^2 \times h = \tfrac{1}{3}a^2h$$

For a cone, see fig. 8.2

$$\text{Curved surface area} = \tfrac{1}{2} \times 2\pi r \times l = \pi r l$$
$$\text{Volume} = \tfrac{1}{3} \times \pi \times r^2 \times h = \tfrac{1}{3}\pi r^2 h$$

For a sphere, see fig. 8.3

$$\text{Surface area} = 4\pi r^2$$
$$\text{Volume} = \tfrac{4}{3}\pi r^3$$

Example 8.1 Find the total surface area of a solid hemisphere of diameter 60 mm.

The total surface area of a solid hemisphere is the curved surface and the area of the base, see fig. 8.4

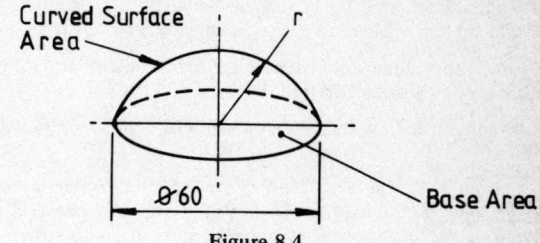

Figure 8.4

$$\text{Curved surface area} = \tfrac{1}{2}(4\pi r^2) = 2\pi r^2$$
$$= 2 \times \pi \times 30^2$$
$$= 5655 \text{ mm}^2$$

$$\text{Area of base} = \pi r^2 = \pi \times 30^2$$
$$= 2827 \text{ mm}^2$$

$$\text{Total surface area} = 8482 \text{ mm}^2$$

Example 8.2 Find the volume of a cone of height 300 mm and base diameter 290 mm.

$$\text{Volume of cone} = \tfrac{1}{3}\pi r^2 h$$
$$= \tfrac{1}{3} \times \pi \times 145^2 \times 300$$
$$= 6\,605\,000 \text{ mm}^3$$

Example 8.3 A solid metal pyramid has a square base of side 250 mm and a vertical height of 160 mm. Calculate the total surface area and also the mass of the pyramid if 1 m³ of the metal has a mass of 7.8×10^3 kg.
In fig. 8.1

$$l^2 = \left(\frac{a}{2}\right)^2 + h^2$$
$$l^2 = 125^2 + 160^2 = 41\,225$$
$$l = 203 \text{ mm}$$

A similar method to the above must always be used to find the slant height when the vertical height is given or to find the vertical height when the slant height is given.
In the pyramid $a = 250$ mm, $h = 160$ mm, $l = 203$ mm.

$$\text{Total surface area} = 2al + a^2$$
$$= 2 \times 250 \times 203 + 250 \times 250$$
$$= 164\,000 \text{ mm}^2$$

$$\text{Volume} = \tfrac{1}{3}a^2 h = \tfrac{1}{3} \times 250 \times 250 \times 160 = 3\,333\,000 \text{ mm}^3$$
$$= 0.003\,333 \text{ m}^3$$

Mass $= 0.003\,333 \times 7.8 \times 10^3 = 26$ kg
Total surface area $= 164\,000$ mm² and mass $= 26$ kg

EXERCISE 8.1

1. Find the volume of a square pyramid with a perpendicular height of 1.2 m and the length of a side of the square of the square base 0.9 m.

2. Find the curved surface area of a cone of perpendicular height 45 mm and base radius 35 mm.

3. Find the volume of a sphere of radius 36 mm.

4. A square pyramid has a vertical height of 75 mm and a base of side 80 mm. Calculate the total surface area of the pyramid.

5. Find the volume of a cone with slant height 100 mm and diameter of the base 120 mm.

6. A sphere has a surface area of 8765 mm². Calculate its radius.

7. A hopper for holding sand is an inverted square pyramid. If the side of the square top is 3.6 m and the perpendicular height is 3.2 m find the volume of sand it holds when full.

8. The curved surface area of a cone is 8792 mm² and the base radius is 40 mm. Find the slant height.

8.2 FRUSTA OF PYRAMIDS AND CONES, CAPS AND ZONES OF SPHERES

If a pyramid or cone is cut by a plane parallel to the base, and the apex part removed, the remaining part is called a frustum (plural of frustum is frusta).
The area of the sloping faces =
$\tfrac{1}{2}$ (sum of top and bottom perimeter) × (the slant height)
The volume $= \tfrac{1}{3}$ of the vertical height of the frustum
$\times (a + A + \sqrt{aA})$ where a = area of top, A = area of base

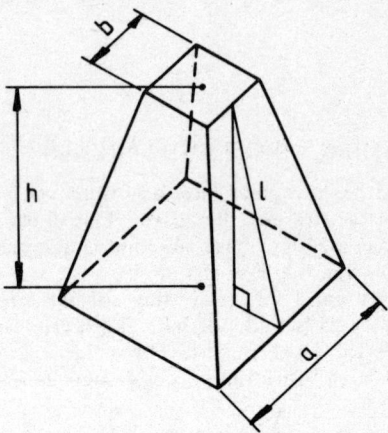

Figure 8.5

For a frustum of a square pyramid, see fig. 8.5
Surface area of sloping sides $= \tfrac{1}{2}(4b + 4a) \times l = 2(b + a)l$

Volume $= \tfrac{1}{3}h(a + A + \sqrt{aA})$, where $a = b^2$, $A = a^2$,
$$\sqrt{aA} = ba$$
$$= \tfrac{1}{3}h(a^2 + b^2 + ba)$$

For a frustum of a cone, see fig. 8.6

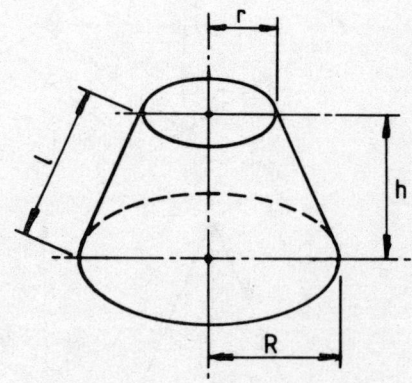

Figure 8.6

Curved surface area $= \tfrac{1}{2}(2\pi r + 2\pi R) \times l$
$$= \pi l(r + R)$$

Volume $= \tfrac{1}{3}h(a + A + \sqrt{aA})$, where $a = \pi r^2$, $A = \pi R^2$,
$$= \tfrac{1}{3}h(\pi r^2 + \pi R^2 + \pi rR)$$
$$= \tfrac{1}{3}\pi h(r^2 + R^2 + rR)$$

When a sphere is cut by a plane, the smaller part is usually called a spherical cap. For a spherical cap, see fig. 8.7

Curved surface area $= 2\pi Rh$, Volume $= \dfrac{\pi h^2}{3}(3R - h)$

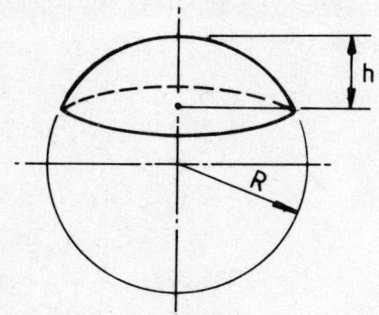

Figure 8.7

When a sphere is cut by two parallel planes, see fig. 8.8, the part between the planes is called a zone. This is treated as the difference between two caps.

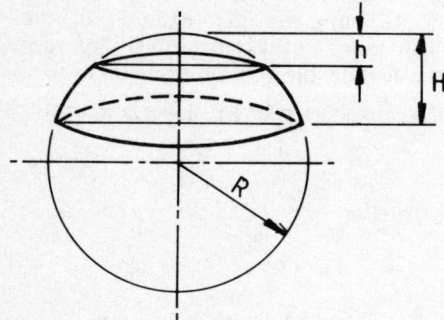

Figure 8.8

Curved surface area $= 2\pi RH - 2\pi Rh = 2\pi R(H - h)$

Volume of zone $= \dfrac{\pi H^2}{3}(3R - H) - \dfrac{\pi h^2}{3}(3R - h)$

Example 8.4 A tapered roller is 120 mm long. Its end radii are 12 mm and 32 mm. Find the curved surface area of the roller. 1 m³ of the material of which the roller is made has a mass of 2.8×10^3 kg, find the mass of the roller.

We must first find the slant height of the roller. In fig. 8.9, using the theorem of Pythagoras

$$AB^2 = BC^2 + CA^2$$
$$AB^2 = 120^2 + (32 - 12)^2$$
$$AB^2 = 14\,400 + 400 = 14\,800$$

$$AB = \sqrt{14\,800} = 121.7 \text{ mm}$$

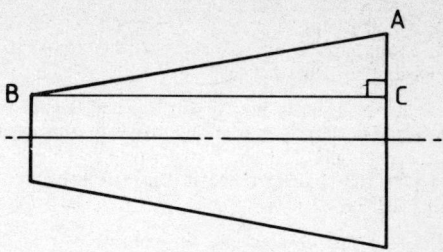

Figure 8.9

Curved surface area of roller $= \pi l(r + R)$

$$= \pi \times 121.7(12 + 32)$$
$$= \pi \times 121.7 \times 44 = 16\,820 \text{ mm}^2$$

Volume of roller $= \tfrac{1}{3}\pi h(r^2 + R^2 + rR)$

$$= \tfrac{1}{3} \times \pi \times 120(12^2 + 32^2 + 12 \times 32)$$
$$= \tfrac{1}{3} \times \pi \times 120 \times 1552$$
$$= 195\,030 \text{ mm}^3$$
$$= 0.000\,195 \text{ m}^3$$

Mass of roller $= 0.000\,195 \times 2.8 \times 10^3$ kg $= 0.546$ kg

Example 8.5 A hopper is a frustum of a square pyramid. The top has a side of 2.26 m, the bottom has a side of 1.42 m, and the perpendicular height of the frustum is 3.24 m.
Assuming the dimensions are the internal dimensions of the hopper calculate the volume it will hold when level full.

The volume of the hopper $= \tfrac{1}{3}h(a^2 + b^2 + ba)$
where $h = 3.24$ m, $a = 2.26$ m, $b = 1.42$ m

Volume $= \tfrac{1}{3} \times 3.24 \times (2.26^2 + 1.42^2 + 2.26 \times 1.42)$
$$= \tfrac{1}{3} \times 3.24 \times 10.333$$
$$= 11.16 \text{ m}^3$$

The hopper will hold 11.16 m³ when level full.

Example 8.6 Find the curved surface area and volume of a spherical cap of height 40 mm which has been cut from a sphere of radius 60 mm.

Curved surface area $= 2\pi Rh$

$$= 2 \times \pi \times 60 \times 40$$
$$= 15\,080 \text{ mm}^2$$

Volume $= \dfrac{\pi h^2}{3}(3R - h)$

$$= \dfrac{\pi \times 40^2}{3}(3 \times 60 - 40)$$

$$= \dfrac{\pi \times 1600}{3} \times 140$$

Volume $= 235\,000 \text{ mm}^3$

EXERCISE 8.2

1. A bearing is a frustum of a cone of height 50 mm. If the end radii are 15 mm and 9 mm calculate (a) the curved surface area, (b) the volume of the bearing.

2. A metal spacing piece is in the form of a frustum of a square pyramid. The top is a square of side 40 mm and the bottom a square of side 52 mm. If the piece is 15 mm thick find the mass of the spacing piece if 1 m³ of the metal has a mass of 7.8×10^3 kg.

3. Find the curved surface area and the volume of a spherical cap of height 25 mm cut from a sphere of radius 45 mm.

4. A frustum of a solid cone has a volume of 516 000 mm³. If the base radius is 70 mm and the top radius is 50 mm, find the height of the frustum.

5. A tapered bar has an hexagonal cross-section and is 300 mm long. At one end the side of the regular hexagon is 45 mm and at the other end it is 25 mm. Calculate the mass of the bar if 1 m³ of the metal has a mass of 2.7×10^3 kg.

6. A metal spacing piece is a zone of a sphere 15 mm thick and is cut from a sphere of radius 60 mm. If the two parallel planes forming the zone are 20 mm and 35 mm from the centre of the circle find the volume of the spacing piece. Find also its mass if 1 m³ of the metal has a mass of 2.7×10^3 kg.

8.3 GENERAL EXAMPLES

Example 8.7 A sphere of diameter 50 mm has a conical hole in it whose apex is at the centre of the sphere and whose vertical angle is 90°. Find the volume of the remaining part of the sphere.

$$\text{Volume remaining} = \text{volume of sphere} -$$
$$(\text{volume of cap} + \text{volume of cone})$$

$$= \tfrac{4}{3}\pi R^3 - \left(\frac{\pi h^2}{3}(3R - h) + \tfrac{1}{3}\pi r^2 H \right) \qquad (1)$$

In fig. 8.10, $R = 25$ mm

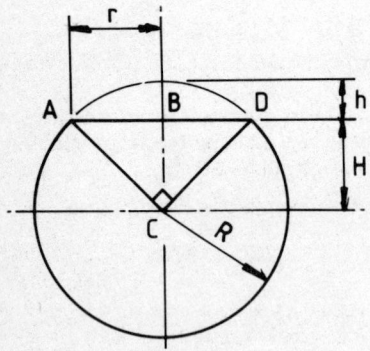

Figure 8.10

In $\triangle ABC$ since $\angle ACD = 90°$
$\angle ACB = \angle BAC$ and hence $r = H$
but $r^2 + H^2 = R^2$
$r^2 + r^2 = R^2 \quad$ or $\quad 2r^2 = R^2$

$R = 25$, hence $2r^2 = 25^2$

or $r^2 = \dfrac{25^2}{2}, r = 17.68$ mm

$r = 17.68$ mm, $H = 17.68$ mm

Also $h = R - H = 25 - 17.68 = 7.32$ mm
Hence required volume from (1) is

$$\text{Volume remaining} = \tfrac{4}{3}\pi \times 25^3$$
$$- \left(\frac{\pi \times 7.32^2}{3}(3 \times 25 - 7.32) \right.$$
$$\left. + \tfrac{1}{3} \times \pi \times 17.68^2 \times 17.68 \right)$$
$$= 65\,449.8 - (56.111 \times 67.68 +$$
$$5787.292)$$
$$= 65\,449.8 - 9584.9$$
$$= 55\,860 \text{ mm}^3$$

The volume remaining $= 55\,860$ mm³

Example 8.8 A metal weight consists of a frustum of a circular cone surmounted at its smaller end by a hemisphere. The end diameters of the frustum are 70 mm and 120 mm and the diameter of the hemisphere is 100 mm. If the total height of the weight is 150 mm calculate the volume of metal in the weight.

$$\text{Volume of hemisphere} = \tfrac{1}{2}(\tfrac{4}{3}\pi r^3) = \tfrac{2}{3} \times \pi \times 50^3$$
$$= 261\,799 \text{ mm}^3$$

$$\text{Volume of frustum} = \frac{\pi h}{3}(r^2 + R^2 + rR)$$

$$h = 150 - 50 = 100 \text{ mm}$$

$$\text{Volume} = \frac{\pi \times 100}{3}(35^2 \times 60^2 + 35 \times 60)$$

$$= \frac{\pi \times 100}{3} \times 6925$$

$$= 725\,184 \text{ mm}^3$$

$$\text{Total volume of weight} = 725\,184 + 261\,799$$
$$= 987\,000 \text{ mm}^3$$

EXERCISE 8.3

1. A bucket, in the shape of a frustum of a cone of end diameters 250 mm and 100 mm and depth 380 mm is filled to the top with water. The water is used to fill a vessel in the shape of a sphere of radius 125 mm. Find the volume of water remaining in the bucket after the sphere is filled.
What would be the radius of the spherical vessel if it just held all the water in the bucket?

2. A rivet consists of 2 hemispheres and a cylinder and has the dimensions shown in fig. 8.11, all dimensions in mm. Calculate its mass if 1 m³ of the metal has a mass of 7.8×10^3 kg.

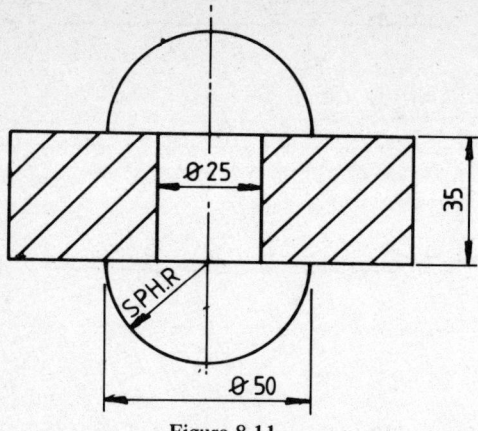

Figure 8.11

3. A casting consists of a frustum of a cone with a cylindrical hole bored in the base, as shown in fig. 8.12, all dimensions in mm. If it is made of metal of density 7.7×10^3 kg/m³ calculate the mass of the casting.

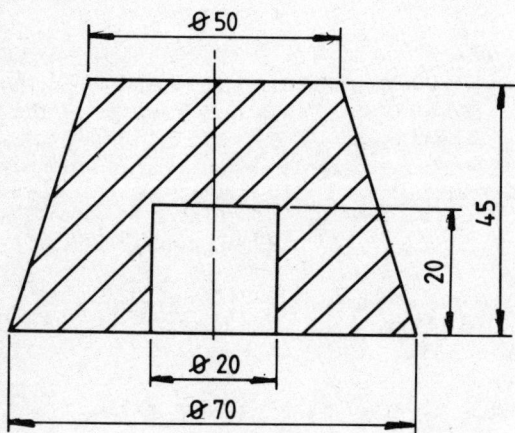

Figure 8.12

4. A tapered pin of length 200 mm and with end radii 25 mm and 35 mm is turned from a cylindrical rod of radius 35 mm. Calculate the volume of metal removed.

5. The section of an extruded rod is shown in fig. 8.13, all dimensions in mm. Find its mass per metre length if the density of the metal is 8.26×10^3 kg/m³.

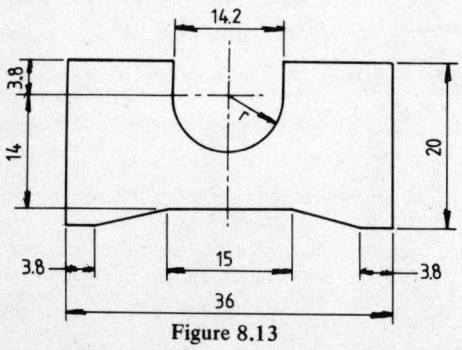

Figure 8.13

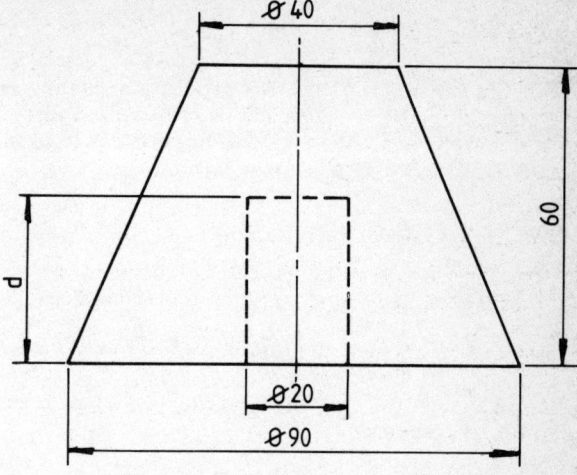

Figure 8.14

6. Fig. 8.14 shows a balance weight in the shape of a frustum of a cone. It is to be lightened by 5% by drilling a 20 mm diameter hole in the base. Calculate the depth, d, to which the hole should be drilled.

SUMMARY

1. Pyramid:
Area of sloping faces $= \frac{1}{2}$(perimeter of base)
$\qquad\qquad\qquad\qquad \times$ (slant height)
Volume $= \frac{1}{3}$(area of base) × (vertical height)

2. Cone:
Curved surface area $= \pi r l$
Volume $= \frac{1}{3}\pi r^2 h$
r = radius of base, l = slant height, h = vertical height.

3. Sphere:
Surface area $= 4\pi R^2$
Volume $= \frac{4}{3}\pi R^3$, where R = radius.

4. When a pyramid or cone is cut by a plane parallel to the base and the apex part is removed the remainder is called a frustum.

5. Frustum of a pyramid:
Area of sloping faces
$= \frac{1}{2}$(sum of top and bottom perimeters) × (slant height)
Volume $= \frac{1}{3}h(a + A + \sqrt{aA})$
h = vertical height, a = area of top, A = area of bottom.

6. Frustum of cone:
Curved surface area $= \pi(r + R)l$

Volume $= \dfrac{\pi h}{3}(r^2 + R^2 + rR)$

r = radius of top, R = radius of bottom, l = slant height, h = vertical height.

7. When a sphere is cut by a plane the smaller part is called a cap. If the depth is h and the radius of the

sphere R then the curved surface area $= 2\pi Rh$, and the volume $= \dfrac{\pi h^2}{3}(3R - h)$.

8. When a sphere is cut by two parallel planes the part between the planes is called a zone. The curved surface area and the volume of a zone can be found by treating it as the difference between two caps.

SELF ASSESSMENT PAPER No 8

Instructions: Answer all questions in both sections
Time allowed   Section A 50 minutes (10 marks each question)
                Section B 40 minutes (20 marks each question)

Marks gained: 50+ pass with credit, 40–50 pass, less than 40 fail, repeat chapter 8

Section A

1. The square base of a pyramid has side of length 2 m. If the height of the pyramid is 1.3 m find its volume and the area of its sloping sides.

2. Find the curved surface area of a cone of slant height 60 mm and base radius 30 mm. Calculate also the volume.

3. Find the volume of a sphere of radius 2.5 m. Also find the surface area.

4. Calculate the mass of a tapered roller 200 mm long with end radii 13 mm and 35 mm if 1 m³ of the metal has a mass of 2.7×10^3 kg.

5. Find the area of the sloping faces of a frustum of a square pyramid of height 25 mm if a side of the top is 10 mm and a side of the bottom is 25 mm.

6. From a solid sphere of radius 100 mm a zone is cut so that the plane faces are 40 mm and 60 mm from the centre of the sphere. Calculate the curved surface area of the zone.

Section B

1. A gearwheel blank consists of a frustum of a cone and a cylinder as shown in fig. 8.15. Calculate its mass if the density of the material from which it is made is 7750 kg/m³.

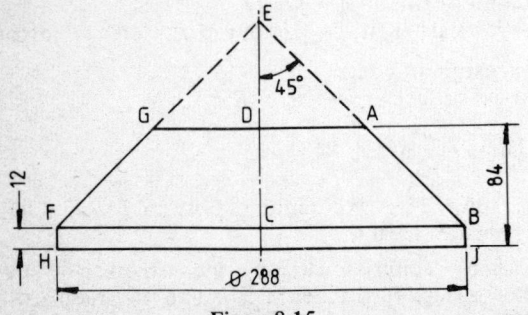

Figure 8.15

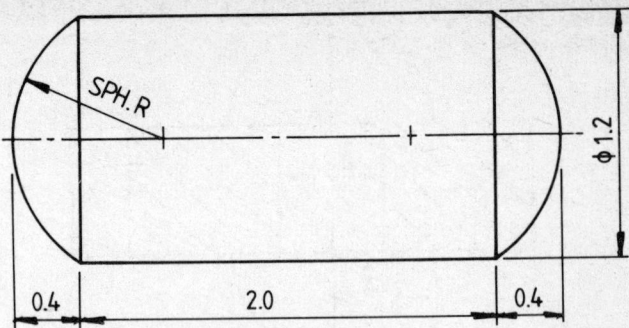

Figure 8.16

2. A fuel tank consists of a cylinder with two spherical caps as shown in fig. 8.16, all dimensions in metres. The spherical ends are parts of a sphere of radius 0.9 m. Calculate, (a) the total surface area of the tank, (b) the volume of the tank.

ANSWERS

Exercise 8.1   1.   0.324 m³,   2.   6268 mm²,
3.   195 000 mm³,   4.   16 000 mm²,
5.   301 600 mm³,   6.   26.4 mm,   7.   13.82 m³,
8.   70.0 mm.

Exercise 8.2   1.   (a) 3800 mm², (b) 23 090 mm³,
2.   0.249 kg,   3.   71 990 mm²,   4.   45.2 mm,
5.   2.65 kg,   6.   133 125 mm², 0.359 kg.

Exercise 8.3   1.   0.001 52 m³, 133 mm,
2.   0.645 kg,   3.   0.953 kg,   4.   199 000 mm³,
5.   4.33 kg,   6.   33.3 mm.

SELF ASSESSMENT PAPER No 8

	Marks
Section A	
1. Volume $= \frac{1}{3} \times 1.3 \times 2^2$	2
Volume $= 1.73$ m³	1
$l^2 = 1.3^2 + 1^2$	2
$l = 1.64$ m	1
Area of sloping sides $= \frac{1}{2} \times 4 \times 2 \times 1.64$	2
Area $= 6.56$ m²	2
2. Curved surface area $= \pi \times 60 \times 30$	2
Area $= 5655$ mm²	1
$h^2 = 60^2 - 30^2$	2
$h = 52$ mm	1
Volume $= \frac{1}{3} \times \pi \times 30^2 \times 52$	2
Volume $= 49\,000$ mm³	2
3. Volume of sphere $= \frac{4}{3} \times \pi \times 2.5^3$	3
$= 65.45$ mm³	2
Surface area of sphere $= 4 \times \pi \times 2.5^2$	3
$= 78.54$ mm²	2
4. Volume $= \frac{1}{3} \times \pi \times 200(35^2 + 13^2 + 13 \times 25)$	3
Volume $= 387\,000$ mm³	3

Marks

$$\text{Mass} = \frac{387\,000}{10^9} \times 2.7 \times 10^3$$ 2

$\text{Mass} = 1.046\,\text{kg}$ 2

5. Slant height $l = \sqrt{25^2 + 7.5^2} = 26.1\,\text{mm}$ 4

Area of sloping faces $= \frac{1}{2}(4 \times 25 + 4 \times 10) \times 26.1$ 4
$$= 1827\,\text{mm}^2$$ 2

6. $H = 60, h = 40$ 4

Curved surface area $= 2\pi \times 100 \times (60 - 40)$ 4
$$= 12\,570\,\text{mm}^2$$ 2

Section B

1. $AD = ED = 72\,\text{mm}, DC = 84 - 12 = 72\,\text{mm}$ 4

Volume of gearwheel blank
  $=$ volume of frustum of cone ABFG $-$
    volume of cylinder BFHJ   2

$$= \frac{72 \times \pi}{3}(144^2 + 72^2 + 144 \times 72)$$

$$+ \pi \times 144^2 \times 12$$ 6

$$= 24\pi \times 36\,294 + 12\pi \times 20\,740$$ 2

$$= 3\,518\,000\,\text{mm}^3$$ 2

Marks

$$\text{Mass} = \frac{3\,518\,000}{10^9} \times 7750$$ 2

$$= 27.3\,\text{kg}$$ 2

2. Total surface area $= 2 \times$ area of cap $+$
                      area of cylinder   2

$$= 2 \times 2\pi \times 0.9 \times 0.4 + 2\pi \times 0.6 \times 2.0$$ 4

$$= 4\pi \times 0.36 + 2\pi \times 1.2$$ 2

$$= 12.06\,\text{m}^2$$ 2

Volume of tank $= 2 \times$ volume of cap
                $+$ volume of cylinder   2

$$= 2 \times \frac{\pi \times 0.4^2}{3}(3 \times 0.9 - 0.4) + \pi \times 0.6^2 \times 2.0$$ 4

$$= \frac{0.32\pi}{3} \times 2.3 + \pi \times 0.72$$ 2

$$= 3.033\,\text{m}^3$$ 2

9 Irregular Areas and Volumes

9.1 THE MID-ORDINATE RULE

The mid-ordinate rule is a method of determining the area of an irregular figure. The figure is divided into strips of equal width and the centre line of each strip is drawn. The lines forming the strips are called ordinates and the centre lines are called the mid-ordinates. We then consider that the length of each mid-ordinate is the average length of the strip and so the area will be the width of the strip times the length of the mid-ordinate.

If the mid-ordinates are of length $l_1, l_2, l_3, \ldots l_n$ and the width of each strip is w then the area of the figure will be approximately $wl_1 + wl_2 + wl_3 + \ldots + wl_n$. This is written area $\approx w(l_1 + l_2 + l_3 + \ldots + l_n)$

Example 9.1 Draw the graph of $y = \sin x$ for values of x from $0°$ to $180°$ (0 to π radians). Use the mid-ordinate rule to find the area between the curve and the x-axis.

x	$0°$	$30°$	$60°$	$90°$	$120°$	$150°$	$180°$
$y = \sin x°$	0	0.5	0.866	1.00	0.866	0.5	0

The graph of $y = \sin x$ is drawn in fig. 9.1.

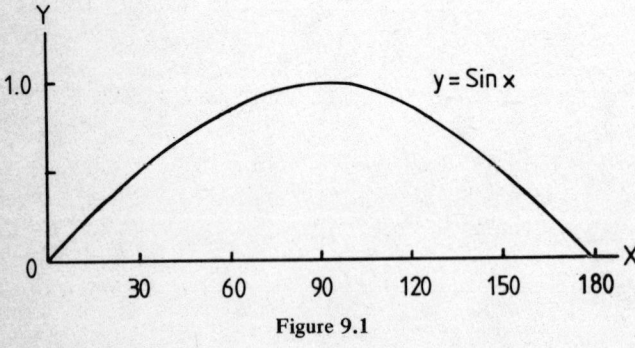

Figure 9.1

The lengths of the mid-ordinates are 0.26, 0.71, 0.97, 0.97, 0.71, 0.26.
The width must be taken in radians, not degrees.

The width of each strip $= 30° = 30 \times \dfrac{\pi}{180} = \dfrac{\pi}{6}$ rad.

Using the mid-ordinate rule

$$\text{Area} = \frac{\pi}{6}(0.26 + 0.71 + 0.97 + 0.97 + 0.71 + 0.26)$$

$$= \frac{\pi}{6} \times 3.88$$

$$= 2.03$$

The true value by calculus is 2.00. A more accurate value could be obtained by taking more strips of smaller width. The mid-ordinate rule is only an approximate way of finding an area of an irregular shape and hence the above result can be assumed to be good.

EXERCISE 9.1

1. The depth of water in a river was measured at every 5 m across the width of the river. The results are shown in the table below.

Distance from river bank (m)	0	5	10	15	20	25	30	35
Depth of water (m)	0	0.6	1.2	1.8	2.9	1.7	0.9	0

Draw a graph to represent these values and use the mid-ordinate rule to find the cross-sectional area of the water in the river.

2. The speed of a train over the first minute of its journey is recorded and the results given below

Speed V (m/s)	0	1.6	3.8	6.1	8.8	11.4	14.6
Time t (s)	0	10	20	30	40	50	60

Draw a graph of V (vertically) against t. Use the mid-ordinate rule to find the area between the curve and the t-axis and so find the total distance travelled in the minute.

9.2 THE TRAPEZOIDAL RULE

The graph is drawn as before and the area divided into strips of equal width, as before. But now each strip is considered as a trapezium. Thus if the lengths of the sides of each strip are $y_1, y_2, y_3, \ldots y_n$ and the width is w the area is approximately

$$\frac{y_1 + y_2}{2} \times w + \frac{y_2 + y_3}{2} \times w \text{ etc.}$$

or Area $= \dfrac{w}{2}(y_1 + y_2 + y_2 + y_3 + y_3 \ldots + y_n)$

$\qquad = \dfrac{w}{2}(y_1 + 2y_2 + 2y_3 \ldots + y_n)$

$\qquad = \dfrac{w}{2}(y_1 + y_n + 2[y_2 + y_3 + \ldots])$

$\qquad = w(\tfrac{1}{2}[y_1 + y_n] + [y_2 + y_3 + \ldots])$

or Area $=$
   width of one strip $\times$ ($\tfrac{1}{2}$[first and last ordinates]
      $+$ the sum of all the other ordinates)

Example 9.2 If we consider again the sine curve from $0°$ to $180°$ we have

x	0	30	60	90	120	150	180
$y = \sin x$	0	0.500	0.866	1.000	0.866	0.500	0

Width of one strip $= \dfrac{\pi}{6}$ radians

Area $= \dfrac{\pi}{6}(\tfrac{1}{2}[0 + 0]$

$\qquad + [0.5 + 0.866 + 1.00 + 0.866 + 0.5])$

$\qquad = \dfrac{\pi}{6}(3.732)$

$\qquad = 1.954$

In this case the area is less than the true value of 2.00. The reason for this is that we have assumed each strip is a trapezium, which in this case is less than the true area, see fig. 9.2. But again the trapezoidal rule is only an approximation and 1.954 is a reasonable approximation to 2.000. A more accurate result can be obtained by taking more strips. It is not always necessary to draw a graph when using the trapezoidal rule since if the ordinates are given at regular intervals they can be used directly.

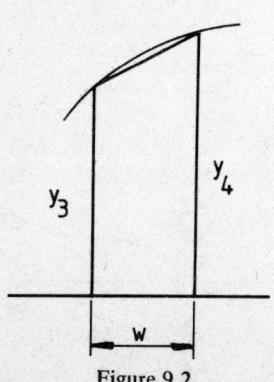

Figure 9.2

Example 9.3 The cross-section of an extruded metal bar is 72 mm wide. The depth of the bar at 8 mm intervals across the bar is given in the table below.

Distance from one side (mm)	0	8	16	24	32	40
Depth of bar (mm)	15.0	22.6	25.1	30.1	23.2	17.0

	48	56	64	72
	16.3	13.8	10.0	9.0

Use the trapezoidal rule to calculate the area of the cross-section. We need not draw a graph, we can apply the trapezoidal rule directly.

Area $\doteqdot 8(\tfrac{1}{2}[15.0 + 9.0] + [22.6 + 25.1 + 30.1 + 23.2$

$\qquad + 17.0 + 16.3 + 13.8 + 10.0])$

$\qquad = 8(12.0 + 158.1) = 8 \times 170.1 = 1360.8$

Area of cross-section $= 1360 \text{ mm}^2$

EXERCISE 9.2

1. It is required to find the cross-sectional area of a ventilation duct. The depth of the duct is measured at various points across the section and the results recorded in the table below:

Distance across duct d (mm)	0	40	85	120	185	235	270	335	400
Depth x (mm)	100	125	140	165	170	164	150	132	110

Draw a graph of x against d and use the trapezoidal rule to determine the cross-sectional area.

2. The work done can be found by finding the area under a force against distance graph. The force required to move a mass over rough ground was measured at half metre intervals.

Force (N)	3.84	4.15	4.72	4.28	4.34	4.21	3.82	3.76	3.54
Distance (m)	0	0.5	1.0	1.5	2.0	2.5	3.0	3.5	4.0

Use the trapezoidal rule to find the work done.

Note: If further examples are required use the two examples in EXERCISE 9.1.

9.3 SIMPSON'S RULE

When using Simpson's rule there must be an even number of strips, that is an odd number of ordinates. If the lengths of the ordinates are $y_1, y_2, y_3, y_4, y_5, y_6, y_7$ and the width of each strip is w then:

Area $= \dfrac{w}{3}(y_1 + y_7 + 4[y_2 + y_4 + y_6] + 2[y_3 + y_5])$

or area $= \dfrac{\text{width of one strip}}{3}$ (first $+$ last ordinates

$+ 4$ times even ordinates $+ 2$ times other odd ordinates)

Example 9.4 If we consider again the sine curve from $0°$ to $180°$ we have

$x°$		30	60	90	120	150	180
$y = \sin x°$		0.500	0.866	1.000	0.866	0.500	0

Width of one strip $= \dfrac{\pi}{6}$ radians

$$\text{Area} = \frac{\frac{\pi}{6}}{3}(0 + 0 + 4[0.500 + 1.000 + 0.500]$$
$$+ 2[0.866 + 0.866])$$

$$\text{Area} = \frac{\pi}{18}(0 + 8.0 + 3.464)$$

$$= \frac{\pi \times 11.464}{18} = 2.001$$

In this example Simpson's rule gives a very good estimate of the area.

In general, Simpson's rule gives a more accurate estimate of the area, but any of the three methods can give a good estimate, more strips being used if a more accurate result is required.

Example 9.5 The pressure and volume of a gas are measured at various intervals as the gas expands. The results are given in the table below.

Pressure (kN/m²)	172		112		80	61		48
Volume (m³)		0.021	0.028	0.035	0.042	0.049

	39	33
	0.056	0.063

Find the work done by the gas.
The work done by the gas as it expands is equal to the area under the pressure–volume curve. Since the values are given at regular intervals we need not draw a graph. Also since we have an even number of strips we can use Simpson's rule.

$$\text{Area} = \frac{0.007}{3}(172 + 33 + 4[112 + 61 + 39]$$

$$+ 2[80 + 48])$$

$$= \frac{0.007}{3}(205 + 848 + 256)$$

$$= \frac{0.007}{3} \times 1309 = 3.054$$

The work done by the gas $= 3.054$ kNm.

EXERCISE 9.3

1. It is required to concrete an irregular shaped area of length 50 m. The widths at various distances from one end are measured and recorded below.

Distance from end (m)	0	6	15	22	29	34	39	45
Width (m)	3.6	4.2	3.9	4.7	5.3	6.1	5.4	5.2

50	60
4.5	4.2

Draw a graph of width against distance from end and use Simpson's rule to estimate the area of concrete required.

2. A cover plate is 240 mm long and its width at regular intervals across the plate are given below.

Distance from end (mm)	0	30	60	90	120	150	180	210	240
Width (mm)	0	96	151	168	162	147	120	87	0

Use Simpson's rule to estimate the area of the cover plate.

Note: If further examples are required the examples in EXERCISES 9.1 and 9.2 can be used.

9.4 VOLUMES

The mid-ordinate rule, trapezoidal rule and Simpson's rule can be used to find volumes if the cross-sectional areas are known at various points along the solid.

Example 9.6 A cooling tower has a circular cross-section, the areas at various distances above the ground are given in the table below.

Height from ground (m)	0	6	12	19	24	31	37	40
Cross-sectional area (m²)	1134	531	284	177	177	254	452	707

Draw a graph of cross-sectional area against height and from it find the volume of the cooling tower.

The graph is drawn in fig. 9.3 and the area of the graph is divided into 8 strips of equal width. Using each method in turn we have:

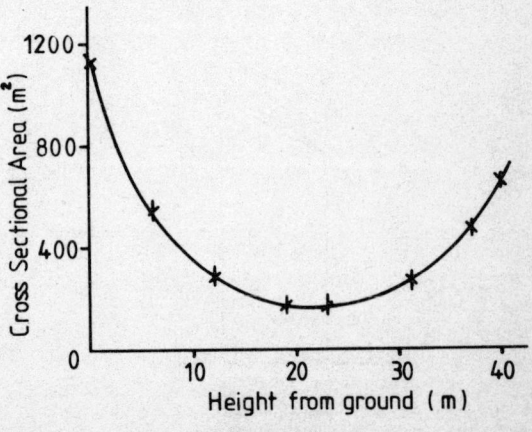

Figure 9.3

(a) Mid-ordinate rule

The mid-ordinates are 860, 440, 270, 180, 180, 200, 300, 480

$$
\begin{aligned}
\text{Volume} &= 5(860 + 440 + 270 + 180 + 180 \\
&\quad + 200 + 300 + 480) \\
&= 5 \times 2910 = 14\,550\,\text{m}^3
\end{aligned}
$$

(b) Trapezoidal rule

The ordinates are 1134, 630, 340, 220, 180, 180, 230, 370, 707

$$
\begin{aligned}
\text{Volume} &= 5(\tfrac{1}{2}[1134 + 707] + 630 + 340 + 220 \\
&\quad + 180 + 180 + 230 + 370) \\
&= 5(920.5 + 2150) = 15\,352\,\text{m}^3
\end{aligned}
$$

(c) Simpson's rule

Using the ordinates from above:

$$
\begin{aligned}
\text{Volume} &= \tfrac{5}{3}(1134 + 707 + 4[630 + 220 + 180 + 370] \\
&\quad + 2[340 + 180 + 230]) \\
&= \tfrac{5}{3}(1841 + 5600 + 1500) = 14\,901\,\text{m}^3
\end{aligned}
$$

In this example the estimates of the volume vary from $14\,600\,\text{m}^3$ to $15\,400\,\text{m}^3$. The Simpson's rule will be the most accurate here, since the curve is always concave the mid-ordinate rule will under estimate the volume and the trapezoidal rule will over estimate the volume.

The volume is approximately $15\,000\,\text{m}^3$

EXERCISE 9.4

1. The cross-sectional area of a casting is measured at intervals throughout its length, the results are shown below.

Distance (mm)	0	5	9	14	20	23	26	30
Area (mm²)	280	290	304	343	427	483	525	588

Draw a graph of the area against the distance and hence determine the volume.

2. A fuel tank is 2 m long and the cross-sectional area is measured at equal distances as follows.

Distance (m)	0	0.25	0.50	0.75	1.00	1.25	1.50
Area (m²)	0	0.260	0.413	0.503	0.503	0.503	0.413

	1.75	2.00
	0.260	0

Calculate the volume of the fuel tank.

3. A forging is 0.90 m long and its area at various distances is recorded below.

Distance (m)	0	0.15	0.30	0.45	0.60
Area of cross-section (m³)	0.090	0.092	0.104	0.122	0.144

	0.75	0.90
	0.171	0.180

Find the volume of the forging. Also calculate the mass of the forging if $1\,\text{m}^3$ of the metal has a mass of $7.8 \times 10^3\,\text{kg}$.

9.5 GENERAL EXAMPLES

EXERCISE 9.5

1. A vehicle starts from rest and its velocity is measured every second.

Time t (s)	0	1.0	2.0	3.0	4.0	5.0	6.0
Velocity v (ms⁻¹)	0	1.3	2.5	3.8	5.3	6.1	9.3

Calculate the distance travelled in the first 6 s, that is the area under the velocity–time graph.

2. The force, F, acting on a body as it moves through a distance, x, from a fixed point is given in the table below.

x (m)	100	200	300	400	450	500	550	600	650
F (N)	350	325	290	260	220	180	130	120	100

Determine the work done by the force in moving through the 650 m.

3. An alternating current I has the following values at equal intervals of 4 ms.

Time (ms)	0	4	8	12	16	20	24	28	32
Current (A)	0	0.8	2.8	4.7	5.3	3.8	2.7	1.2	0

Estimate the average current over the 32 ms period.

4. A ventilation duct is 10 m long and has a varying circular cross-section as shown below.

Distance from one end (m)	0	1	2	3	4	5	6	7	8	9	10
Diameter of duct (m)	1.2	1.1	0.8	0.9	0.9	1.1	1.2	1.1	0.9	0.8	0.8

Estimate the volume of air in the 10 m duct.

SUMMARY

1. If an irregular area is divided into strips of equal width w and the mid-ordinates of the strips are y_1, $y_2, y_3, \ldots y_n$ then the area is given by:

$$\text{Area} = w(y_1 + y_2 + y_3 \ldots + y_n)$$

2. If an irregular area is divided into strips of equal width w and the ordinates are $y_1, y_2, y_3, \ldots y_n$ then the area is given by:

$$\text{Area} = w(\tfrac{1}{2}[y_1 + y_n] + y_2 + y_3 + \ldots)$$

3. If an irregular area is divided into an even number of strips of equal width w then the area is given by:

$$\text{Area} = \frac{w}{3}(y_1 + y_7 + 4(y_2 + y_4 + y_6) + 2[y_3 + y_5])$$

where $y_1, y_2, y_3 \ldots y_7$ are the ordinates.

4. The above methods can be used to estimate volumes if the areas of cross-section are given at regular intervals.

SELF ASSESSMENT PAPER No 9

Instructions: Answer all questions
Time allowed: 45 minutes (question 1, 40 marks,
                question 2, 20 marks)
Marks gained: 30+ pass with credit, 24—30 pass, less
than 24 fail, repeat chapter 9.

1. A template has a horizontal base of 140 mm. The
vertical height at various intervals is recorded below.

Distance (mm)   0   20   35   50   75   85   95   120

Height (mm)   35.5 35.0 33.5 31.0 25.0 18.0 14.0 10.0

                130   140

                2.5   2.0

Draw a graph of height against distance and estimate
the area by using:
(a) the mid-ordinate rule
(b) the trapezoidal rule
(c) Simpson's rule

2. The cross-sectional area of an irregularly shaped
container at regular heights from the base is given
below.

Height (m)   0   0.2   0.4   0.6   0.8   1.0   1.2
Area (m²)   0.44 0.73   0.95   1.14   1.26   1.31   1.33

Estimate the volume of the container by using:
(a) the trapezoidal rule,
(b) Simpson's rule.

ANSWERS

Exercise 9.1   1.   46.0 m²,   2.   396 m

Exercise 9.2   1.   56 900 mm²,   2.   16.5 Nm

Exercise 9.3   1.   279 m²,   2.   28 600 mm²

Exercise 9.4   1.   11 600 mm³   2   0.730 m³,
3.   0.1153 m³, 899 kg

Exercise 9.5   1.   23.2 m   2.   175 700,
3.   2.65 A,   4.   7.80

SELF ASSESSMENT PAPER No 9

	Marks
1. Graph	6

(a) Mid-ordinates 35.5, 34.0, 30.2, 24.5, 16.0,
     8.0, 2.8 mm   3
Width of strip = 20 mm   1
Area = 20(35.5 + 34.0 + 30.2 + 24.5
      + 16.0 + 8.0 + 2.8)   4
Area = 3020 mm²   2

(b) Ordinates 35.5,   35.0,   32.5,   27.8,   20.2,
     12.0,   5.0,   2.0 mm   3
Width of strip = 20 mm   1
Area = 20($\frac{1}{2}$[35.5 + 2.0] + 35.0 + 32.5
      + 27.8 + 20.2 + 12.0 + 5.0)   4
Area = 3025 mm²   2

(c) Ordinates 35.5,   35.5,   35.0,   34.0,   32.5,
     30.2,   27.8,   24.5,   20.2,   16.0,   12.0,
     8.0,   5.0,   2.8,   2.0 mm   5
Width of strip = 10   1

Area = $\frac{10}{3}$(35.5 + 2.0 + 4[35.5 + 34.0
      + 30.2 + 24.5 + 16.0 + 8.0 + 2.8]
      + 2[35.0 + 32.5 + 27.8 + 20.2
      + 12.0 + 5.0])   4

    = $\frac{10}{3}$(37.5 + 604.0 + 265.0)   2

    = 3022 mm²   2

2.(a)
Volume = 0.2($\frac{1}{2}$[0.44 + 1.33] + 0.73 + 0.95
       + 1.14 + 1.26 + 1.31)   7
     = 1.255 m³   3

(b)Volume = $\frac{0.2}{3}$ ([0.44 + 1.33]
       + 4[0.73 + 1.14 + 1.31]
       + 2[0.95 + 1.26])   7

     = 1.261 m³   3

10 Ellipse, Prismoidal Rule, Theorems of Pappus

10.1 THE ELLIPSE

If a cone is cut by a plane which is not parallel to the base it forms an ellipse. An ellipse has two axes of symmetry, the major axis and the minor axis. These axes are shown in fig. 10.1, the major axis is of length $2a$ and the minor axis is of length $2b$.

Area of an ellipse $= \pi ab$
Perimeter of an ellipse $\approx \pi(a + b)$
There is not a simple formula for the exact perimeter of an ellipse.

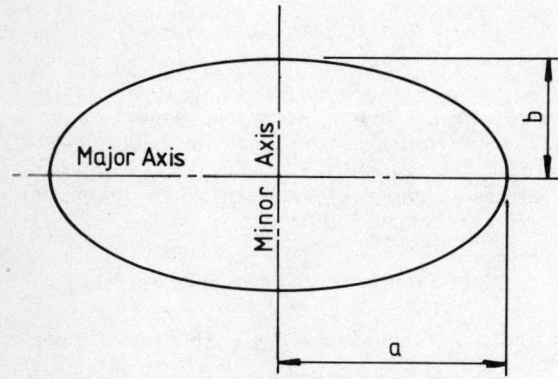

Figure 10.1

Example 10.1 An elliptical cover plate has a major axis of length 112.0 mm and a minor axis of length 70.0 mm. Find the perimeter of the plate and the area. The plate is 12.0 mm thick, find its mass if the metal has a density of 7.8 g/cm³.

$2a = 112$ or $a = 56$ mm, $2b = 70$ or $b = 35$ mm

Perimeter $\approx \pi(a + b) = \pi(56.0 + 35.0) = 285.9$ mm

Area $= \pi ab = \pi \times 56.0 \times 35.0 = 6157.5$ mm²

Volume $=$ Area $\times$ 12.0 $= 6157.5 \times 12.0 = 73\,890$ mm³

Mass $= \dfrac{73\,890}{1000} \times 7.8 = 576$ g

Example 10.2 A solid cylindrical metal bar of diameter 50 mm is cut by a plane making an angle of 40° with the axis of the bar. Calculate the area and the perimeter of the section.

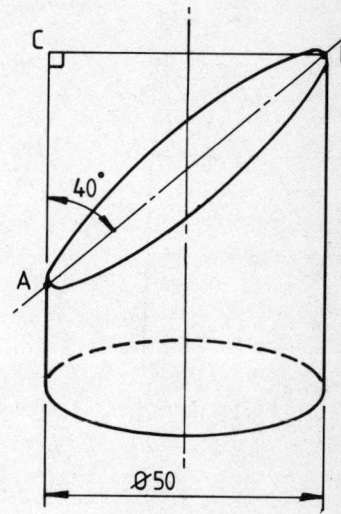

Figure 10.2

In fig. 10.2 the major axis of the ellipse AB can be found from △ABC

$$\sin 40° = \frac{BC}{AB} \quad \text{or} \quad AB = \frac{50}{\sin 40°}$$

$AB = 77.79$ mm

The minor axis will be the diameter of the bar 50 mm.

Hence $a = \dfrac{77.79}{2} = 38.90$ and $b = 25$ mm

Area of section $= \pi ab = \pi \times 38.90 \times 25 = 3055$ mm²
Perimeter of section $\approx \pi(a + b) = \pi \times (38.90 + 25.0)$
$= 201$ mm

EXERCISE 10.1

1. An ellipse has a major axis of 26 mm and a minor axis of 14 mm. Calculate the area and the perimeter of the ellipse.

2. An elliptical cover plate has a major axis of 120 mm and a minor axis of 90 mm, and is 11.0 mm thick. Calculate the perimeter of the plate, the area of the plate and its mass if the metal has a density of 2.8 g/cm³.

3. An ellipse has a major axis $2x$ millimetres long and a minor axis x millimetres long. The area is 8250 mm^2, find its perimeter.

4. A solid cylindrical bar of diameter 30 mm is cut by a plane making an angle of 60° with the axis of the bar. Calculate the area and the perimeter of the resulting section.

10.2 THE PRISMOIDAL RULE

A prismoid is a solid with two parallel faces. The area of the top of the prismoid is a_T, the area of the centre cross-section is a_M and the area of the bottom is a_B. The height of the prismoid is h, see fig. 10.3. We can now use Simpson's rule to find the volume.

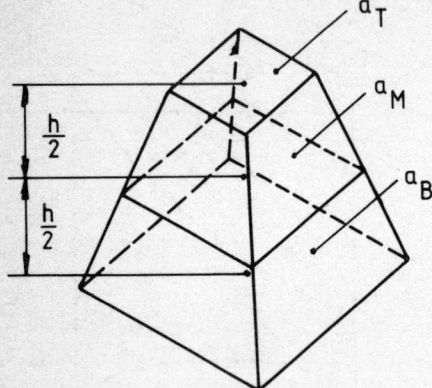

Figure 10.3

$$\text{Volume} = \frac{\frac{h}{2}}{3}(a_T + a_B + 4a_M) = \frac{h}{6}(a_T + a_B + 4a_M)$$

This is known as the prismoidal rule, when the dimensions increase or decrease regularly throughout the prismoid this gives the exact volume. For other prismoids it is an approximation.

Example 10.3 Use the prismoidal rule to find the volume of a frustum of a cone of height 40 mm and end diameters 35 mm and 25 mm.
The diameter at the centre will be the average of the end diameters.

$$d = \frac{25 + 35}{2} = 30$$

$$a_T = \frac{\pi \times 25^2}{4} = 491 \text{ mm}^2, \quad a_B = \frac{\pi \times 35^2}{4} = 962 \text{ mm}^2$$

$$a_M = \frac{\pi \times 30^2}{4} = 707 \text{ mm}^2$$

$$\text{Volume} = \frac{h}{6}(a_T + a_B + 4a_M)$$

$$= \frac{40}{6}(491 + 962 + 4 \times 707)$$

$$\text{Volume} = \frac{40}{6} \times 4281 = 28\,540 \text{ mm}^3$$

The volume of the frustum of the cone = 28 540 mm^3

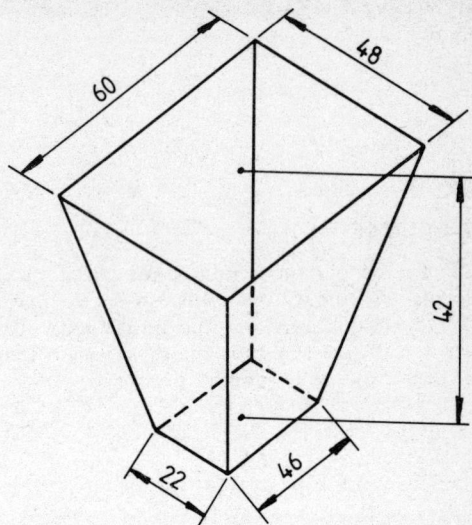

Figure 10.4

Example 10.4 A metal block has parallel rectangular ends as shown in fig. 10.4. The dimensions are in mm. Calculate the volume of the block and its mass if the metal has a density of 7.8 g/cm^3. The dimensions of the centre rectangle will be

$$\frac{46 + 60}{2} = 53 \text{ mm} \quad \text{and} \quad \frac{22 + 48}{2} = 35 \text{ mm}$$

$$a_T = 46 \times 22 = 1012 \text{ mm}^2, \quad a_B = 48 \times 60 = 2880 \text{ mm}^2$$

$$a_M = 53 \times 35 = 1855 \text{ mm}^2$$

$$\text{Volume} = \frac{h}{6}(a_T + a_B + 4a_M)$$

$$= \frac{42}{6}(1012 + 2880 + 4 \times 1855)$$

$$= \frac{42}{6} \times 11\,312 = 79\,184 \text{ mm}^3$$

$$\text{Mass} = \frac{79\,184}{1000} \times 7.8 = 618 \text{ g} = 0.618 \text{ kg}$$

The metal block has a volume of 79 184 mm^3 and a mass of 0.618 kg.

EXERCISE 10.2

1. A frustum of a cone has a height 50 mm and end radii 28 mm and 40 mm. Use the prismoidal rule to find the volume.

2. A hopper has a rectangular top 2.3 m by 1.8 m and a square bottom of side 1.0 m. The perpendicular height of the hopper

is 2.4 m. Calculate the volume of sand the hopper will hold when level full.

3. A solid steel bar is of length 0.25 m and is circular at one end with a radius of 21 mm. The other end is an ellipse with major axis 36 mm and minor axis 24 mm. Calculate the volume of the bar and its mass if the steel has a density of 7.9 g/cm³.

4. A brass block is 36 mm long and has rectangular ends 24 mm by 30 mm and 30 mm by 36 mm. It has a cylindrical hole of diameter 15 mm machined through its centre, from end to end. Calculate the volume of the block and its mass if the density of the brass is 8.3 g/cm³.

10.3 THEOREMS OF PAPPUS

The theorems of Pappus are used to find the surface area and volume of a volume of revolution. These are volumes formed by rotating an area or curve about an axis. For example, if the circle in fig. 10.5 is rotated through 360°, about the $X-X$ axis, it forms a ring. Also if the triangle in fig. 10.6 is rotated through 360°, about the $X-X$ axis it forms a cone.

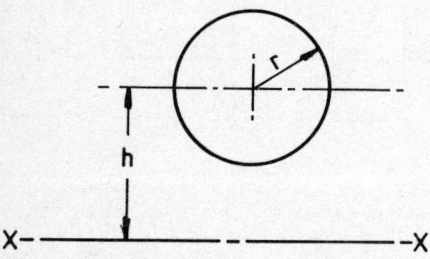

Figure 10.5

Theorem 1. If a curve, which does not cut the $X-X$ axis is rotated about that axis it generates a surface whose area is equal to the product of the length of the curve and the distance travelled by its centre of gravity.

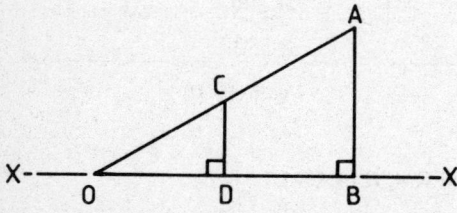

Figure 10.6

Example 10.5 In fig. 10.6 if OA = l and AB = r use the theorem of Pappus to find the curved surface area of a cone of slant height l and base radius r.
In this case the curve is a straight line OA of length l.
If C is the centre of OA then CD will be half of AB.

$$CD = \frac{r}{2}$$

Hence when this line is rotated about the axis $X-X$ the centre of gravity C will move round a circle of radius $\frac{r}{2}$.

Distance moved by centre of gravity C $= 2 \times \pi \times \frac{r}{2} = \pi r$

Length of curve = l
Hence surface area = length of curve × distance moved by its centre of gravity
$$= l \times \pi r$$
$$= \pi r l \text{ (curved surface area of a cone)}$$

Theorem 2. If an area, which does not cut the $X-X$ axis, is rotated about that axis, the volume generated is the product of the area and the distance travelled by its centroid of area.

Example 10.6 In fig. 10.6 if OB = h and AB = r use the theorem of Pappus to find the volume of a cone.
The area of the triangle OAB = $\frac{1}{2}r \times h = \frac{1}{2}rh$
The centroid of a triangle is one third of the height of the triangle from its base.
Hence centroid of $\triangle$OAB is $\frac{1}{3}r$ from base OB.
This centroid will travel round a circle of radius $\frac{1}{3}r$ as the triangle is rotated. Hence distance travelled by
$$\text{centroid} = 2\pi \times \tfrac{1}{3}r = \frac{2\pi r}{3}$$

Volume = Area of $\triangle$AOB × distance travelled by its centroid
$$= \frac{rh}{2} \times \frac{2\pi r}{3} = \tfrac{1}{3}\pi r^2 h \text{ (volume of a cone)}$$

Example 10.7 Find the volume of the ring formed when the circle in fig. 10.5 is rotated about the $X-X$ axis.

Volume = area × distance travelled by its centroid
$$= \pi r^2 \times 2\pi h = 2\pi^2 r^2 h$$

This is the volume of a ring of average radius h and with a circular cross-section of radius r.

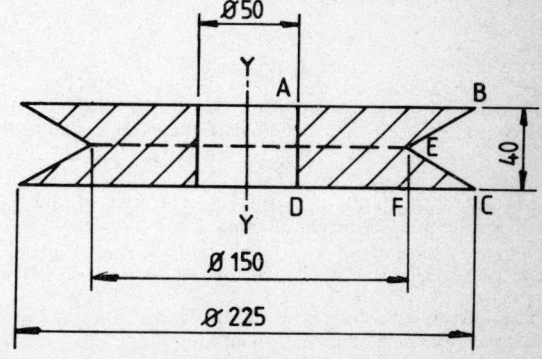

Figure 10.7

Example 10.8 Calculate the mass of the pulley, shown in fig. 10.7. The density of the metal from which it is made is 2720 kg/m³.

This example is solved by considering the rotation of the rectangle ABCD about the $Y-Y$ axis and subtracting the volume formed by rotating the $\triangle$BEC about the axis. Consider the rectangle ABCD, the centroid is $YA + \frac{1}{2}AB$ from the $Y-Y$ axis

$$YA + \tfrac{1}{2}AB = 25 + \tfrac{1}{2}\left(\frac{225-50}{2}\right) = 68.75 \text{ mm}$$

The area ABCD $= AB \times AD = \dfrac{175}{2} \times 40 = 3500 \text{ mm}^2$

Applying the theorem of Pappus

Volume = area × distance travelled by centroid

$$= 3500 \times (2\pi \times 68.75) = 1\,510\,000 \text{ mm}^3$$

Consider $\triangle$BEC, the centroid is $\frac{1}{3}$FC from BC

Distance of centroid from $Y-Y = YC - \frac{1}{3}FC$

$$= \frac{225}{2} - \tfrac{1}{3}\left(\frac{225-150}{2}\right) = 100 \text{ mm}$$

Area of $\triangle$BEC $= \frac{1}{2} \times BC \times FC$

$$= \tfrac{1}{2} \times 40 \times \left(\frac{225-150}{2}\right)$$

$$= 750 \text{ mm}^2$$

Applying the theorem of Pappus

Volume = area × distance travelled by centroid

$$= 750 \times (2\pi \times 100) = 470\,000 \text{ mm}^3$$

Volume of pulley $= 1\,510\,000 - 470\,000$

$$= 1\,040\,000 \text{ mm}^3$$

$$= \frac{1\,040\,000}{10^9} \text{ m}^3$$

$$= 0.001\,04 \text{ m}^3$$

Mass of pulley $= 0.001\,04 \times 2720$

$$= 2.83 \text{ kg}$$

The pulley has a mass of 2.83 kg

EXERCISE 10.3

1. A ring has an average diameter of 30 mm and a circular cross-section of radius 5 mm. Calculate the surface area of the ring.

2. Find the surface area of a frustum of a cone of end radii 40 mm and 30 mm and height 50 mm.

3. The centroid of a semi-circular area of radius r is $\dfrac{4r}{3\pi}$ from the base of the semi-circle. Use the theorem of Pappus to find the formula for the volume of a sphere.

4. Calculate the mass of the light alloy V-pulley shown in fig. 10.8, all dimensions in mm. The mass of 1 m³ of the alloy is 2450 kg.

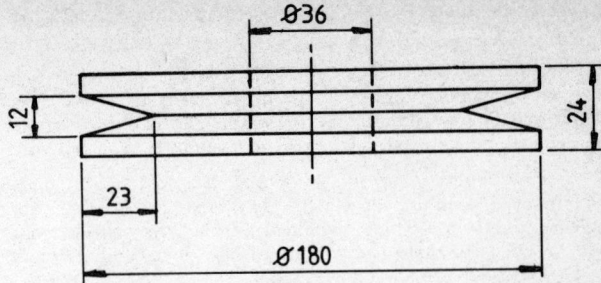

Figure 10.8

5. Fig. 10.9 shows the cross-section of a steel gear blank. Calculate its mass given that the density of steel is 7830 kg/m³, all dimensions in mm.

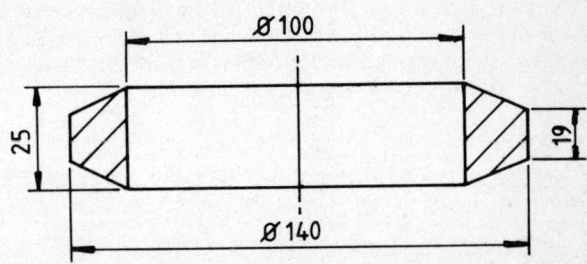

Figure 10.9

6. (a) Calculate the volume of the rubber sealing ring shown in fig. 10.10, all dimensions in mm.
(b) If the mass of the ring is 2.04 kg, calculate the density of the rubber in kg/m³.

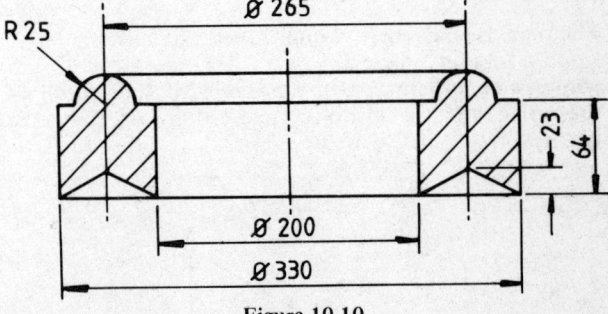

Figure 10.10

10.4 GENERAL EXAMPLES

EXERCISE 10.4

1. A ventilation duct has an elliptical cross-section at one end of major axis 1.04 m and minor axis 0.76 m. The duct reduces over its length of 1.47 m to a circle of radius 0.3 m at the other end. Calculate the volume of air in this duct.

2. A forging is in the form of a ring, of circular cross-section with an internal diameter of 51 mm, and external diameter of 76 mm. By applying Pappus' theorem find the volume and hence the mass, given that the density of the material is 7.0×10^3 kg/m³.

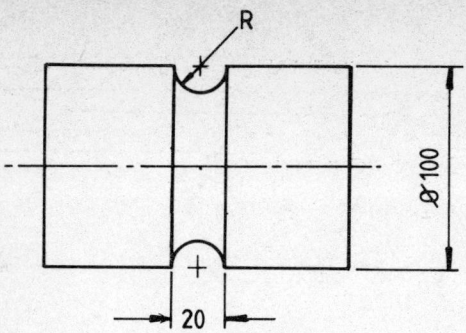

Figure 10.11

3. A semi-circular groove of 20 mm diameter is machined from a shaft of diameter 100 mm. Calculate the volume of metal removed, see fig. 10.11. The centroid of a semi-circle of radius r is $\dfrac{4r}{3\pi}$ from the base.

SUMMARY

1. An ellipse of major axis $2a$, minor axis $2b$ has area πab and approximate perimeter $\pi(a + b)$.

2. A prismoid is a solid with two end faces parallel. If the area of the top is a_T, and the area of the base is a_B, and the area of the middle section is a_M then

volume $= \dfrac{h}{6}(a_T + a_B + 4a_M)$ where h is the height of the prismoid.

3. The first theorem of Pappus states that when a curve is rotated through 360° about an axis:

surface area = length of curve × distance moved by its centre of gravity

4. The second theorem of Pappus states that when an area is rotated through 360° about an axis:

volume generated = area × distance moved by its centroid.

SELF ASSESSMENT PAPER No 10

Instructions: Answer all questions in both sections
Time allowed: Section A 30 minutes
(10 marks each question)
Section B 40 minutes)
(20 marks each question)

Marks gained: 40+ pass with credit, 32–40 pass, less than 32 fail, repeat chapter 10

Section A

1. An ellipse has a major axis of 1.60 m and a minor axis of 0.96 m. Find its area and its perimeter.

2. The sides of the ends of the frustum of a square pyramid are 40 mm and 20 mm. Use the prismoidal rule to find the volume of the frustum, if it has a height of 50 mm.

In Fig. 10.12 the rectangle ABCD is rotated through 360° about the axis $Y-Y$, all dimensions in mm. Use the theorems of Pappus to:
3. Calculate the surface area of the solid formed,

4. Calculate the volume of the solid formed.

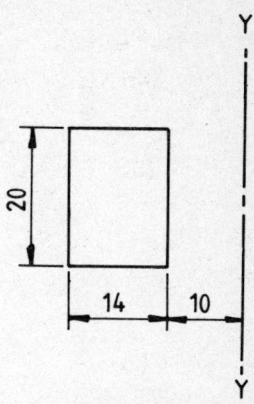

Figure 10.12

Section B

1. A brass block has parallel top and bottom faces. The top is an ellipse of major axis 60 mm and minor axis 40 mm. The bottom is a circle of 80 mm diameter. If the height of the block is 50 mm calculate its mass. The density of the brass is 8.4 g/cm³.

2. Find the mass of the casting shown in fig. 10.13, all dimensions in mm, if the metal has a uniform thickness of 50 mm and a density of 7850 kg/m³.

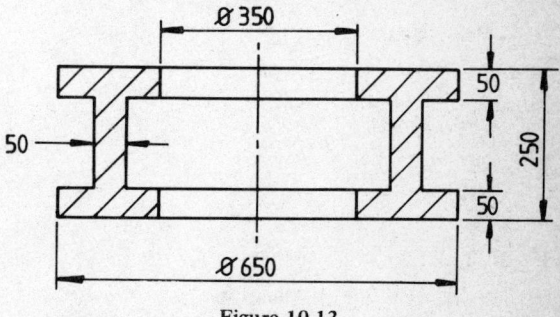

Figure 10.13

ANSWERS

Exercise 10.1 1. 286 mm², 62.8 mm,
2. 330 mm, 8480 mm², 261 g, 3. 342 mm,
4. 816 mm², 102 mm.

Exercise 10.2 1. 83 000 mm³, 2. 5.75 m³,
3. 254 000 mm³, 2.01 kg, 4. 214 g

Exercise 10.3 1. 2961 mm², 2. 10 800 mm²,
3. $\frac{4}{3}\pi r^3$, 4. 1.2 kg, 5. 1.29 kg,
6. (a) 0.003 65 m³, (b) 559 kg/m³

SELF ASSESSMENT PAPER No 10

Marks

Section A

1. $a = 0.8$ m, $b = 0.48$ m   2
Area $= \pi \times 0.80 \times 0.48 = 1.206$ 4
Perimeter $= \pi(0.80 \times 0.48) = 1.206$ 4

2. $a_T = 20 \times 20 = 400 \text{ m}^2$,
$a_B = 40 \times 40 = 1600 \text{ m}^2$ 2
$a_M = 30 \times 30 = 900 \text{ mm}^3$ 2
Volume $= \dfrac{50}{6}(400 + 1600 + 4 \times 900)$ 4
$= 46\,660 \text{ mm}^3$ 2

3. Surface area $= (2 \times (20 + 14)) \times (2\pi \times 17)$ 8
$= 7263 \text{ mm}^2$ 2

4. Volume $= (20 \times 14) \times (2\pi \times 17)$ 8
$= 29\,910 \text{ mm}^3$ 2

Section B

1. $a_T = \pi \times 30 \times 20 = 600\pi$ 2
$a_B = \pi \times 40^2 = 1600\pi$ 2
$a_M = \pi \times 35 \times 30 = 1050\pi$ 4
$h = 50$ 1
Volume $= \dfrac{50}{6}(600\pi + 1600\pi + 4 \times 1050\pi)$ 6
Volume $= 167\,600 \text{ mm}^3$ 2
Mass $= \dfrac{167\,600}{1000} \times 8.4$ g   2
Mass $= 1.41$ kg   1

2. Area $= 2 \times (150 \times 50) + 150 \times 50$ 4
Area $= 22\,500 \text{ mm}^2$ 2
Distance to centroid $= \dfrac{650}{2} - \dfrac{150}{2} = 250$ mm   4
Volume $= 22\,500 \times 2\pi \times 250$ 4
Volume $= 35\,343\,000 \text{ mm}^3$ 2
Mass $= \dfrac{35\,343\,000}{10^9} \times 7850$ 3
Mass $= 277$ kg   1

PHASE TEST 3

Instructions. Answer all questions
Time allowed: One hour (20 marks each question)
Marks gained: 30+ pass with credit, 24–30 pass, less than 24 fail, repeat chapters 8, 9 and 10.

1. A sheet metal bucket is in the shape of a frustum of a cone. The inside diameters of the top and bottom are 360 mm and 270 mm respectively and the depth of the bucket is 360 mm. Find (i) the area of sheet metal required to form the curved surface of the bucket (neglect overlapping at the seams), (ii) the volume of water the bucket would hold.

2. (a) For the wave form shown in fig. 1 determine the mean value.

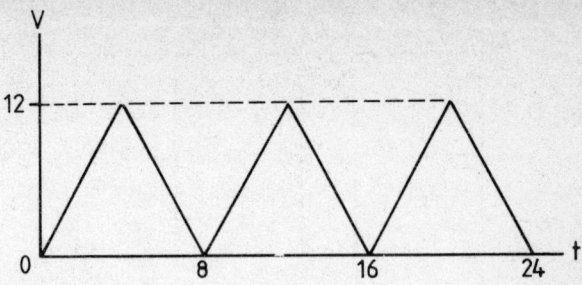

Figure 1

(b) Plot a graph of V^2 over a half cycle and use the mid-ordinate rule to find the area under this graph.

3. (a) A steel barrel is 850 mm high and the diameter at both top and bottom is 600 mm. The diameter midway between the ends is 780 mm. Use the prismoidal rule to calculate the capacity of the barrel in litres. (take 1 litre $= 1000 \text{ cm}^3$).
(b) Use the theorem of Pappus to calculate the mass of the circular pipe shown in fig. 2, all dimensions in mm. The density of the material is $7.5 \times 10^3 \text{ kg/m}^3$.

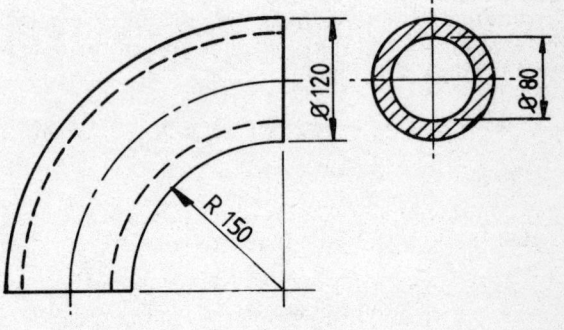

Figure 2

ANSWERS PHASE TEST 3

Marks

1. (i) slant height $l = \sqrt{360^2 + 45^2} = 362.8$ mm   4
Curved surface area $= \pi \times 362.8 \times (180 + 135)$ 6
$= 359\,000 \text{ mm}^2$ 2
(ii) Volume $= \frac{1}{3}\pi \times 360(180^2 + 135^2 + 180 \times 135)$ 5
$= 28\,2000\,000 \text{ mm}^3$ 3
2. (a) Area of triangle $= \frac{1}{2} \times 8 \times 12 = 48$ 4
Mean value $= \dfrac{48}{8} = 6$ 3

(b)

t	0	0.5	1.0	1.5	2.0	2.5	3.0	3.5	4.0
V	0	1.5	3.0	4.5	6.0	7.5	9.0	10.5	12.0
V^2	0	2.25	9.00	20.25	36.00	56.25	81.00	110.25	144.00

7
Area $= 1.0(2.25 + 20.25 + 56.25 + 110.25)$ 4
$= 189$ 2

Marks

3. (a) Volume $= \dfrac{850}{6} (\pi \times 300^2 +$ 4

$\qquad 4 \times \pi \times 390^2 + \pi \times 300^2)$ 4

Volume of barrel $= 351$ litres   3

Marks

(b) Volume $= \frac{1}{4}(\pi \times 120^2 - \pi \times 80^2)$

$\qquad \times \dfrac{\pi \times 210}{2}$ 6

$\qquad = 2\,073\,000 \text{ mm}^3$ 3

$\text{Mass} = \dfrac{2.073 \times 10^6}{10^9} \times 7.5 \times 10^3$ 3

$\qquad = 15.54 \text{ kg}$ 1

11 The Cosine Rule and Area of a Triangle

11.1 THE COSINE RULE

Consider any triangle ABC with base BC and height h.

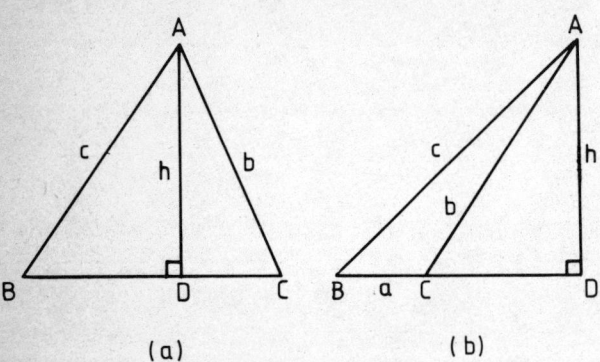

Figure 11.1

In fig. 11.1 (a) $CD = b \cos C$ and $BC = a$ hence $BD = a - b \cos C$

Using the theorem of Pythagoras, in $\triangle ABD$

$$c^2 = h^2 + (a - b \cos C)^2 \qquad (1)$$

and in $\triangle ACD$

$$b^2 = h^2 + (b \cos C)^2 \qquad (2)$$

Take (2) from (1)

$$c^2 - b^2 = (a - b \cos C)^2 - (b \cos C)^2$$

or $c^2 - b^2 = a^2 - 2ab \cos C + b^2 \cos^2 C - b^2 \cos^2 C$

or $\qquad c^2 = a^2 + b^2 - 2ab \cos C \qquad (3)$

In fig. 11.1 (b), $CD = b \cos \angle ACD$
but $\angle ACD = 180° - C$ and therefore
$CD = b \cos (180° - C) = -b \cos C$

$$CD = -b \cos C \quad \text{and} \quad BD = a - b \cos C$$

Using the theorem of Pythagoras, in $\triangle ABD$

$$c^2 = h^2 + (a - b \cos C)^2 \qquad (4)$$

and in $\triangle ACD$

$$b^2 = h^2 + (-b \cos C)^2 \qquad (5)$$

Now equations (4) and (5) are identical to equations (1) and (2) and hence (3) follows.

$$c^2 = a^2 + b^2 - 2ab \cos C$$

is called the cosine rule. By changing the sides around we also have $a^2 = b^2 + c^2 - 2bc \cos A$ and $b^2 = c^2 + a^2 - 2ac \cos B$. The cosine rule is used to solve triangles as follows:

(a) When three sides are given, the cosine rule is used to find the largest angle, the sine rule is then used to find another angle. The sine of an angle between 90° and 180° is positive whereas the cosine is negative. Hence the largest angle is found by the cosine rule since then it is immediately clear whether the angle is greater or less than 90°.

(b) When two sides and the included angle are given, the cosine rule is used to find the third side, the sine rule is then used to find another angle.

Example 11.1 A mass hangs from the junction of two ropes 4.4 m and 3.6 m long respectively. If the other ends of the ropes are attached to the underside of a horizontal beam at points 5 m apart, calculate the angle between the ropes.

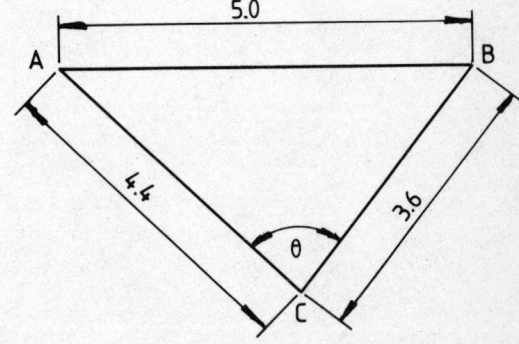

Figure 11.2

Applying the cosine rule to $\triangle ABC$, in fig. 11.2

$$c^2 = a^2 + b^2 - 2ab \cos C$$
$$5^2 = 3.6^2 + 4.4^2 - 2 \times 3.6 \times 4.4 \times \cos \theta$$
$$25 = 12.96 + 19.36 - 2 \times 3.6 \times 4.4 \times \cos \theta$$
$$25 = 32.32 - 31.68 \cos \theta$$
$$31.68 \cos \theta = 32.32 - 25$$

$$\cos \theta = \frac{7.32}{31.68} = 0.2311$$

$$\theta = 76.64° \ (76°38')$$

The angle between the ropes is $76°38'$

Example 11.2 Calculate the centre distance between the holes A and C in fig. 11.3, all dimensions in mm.

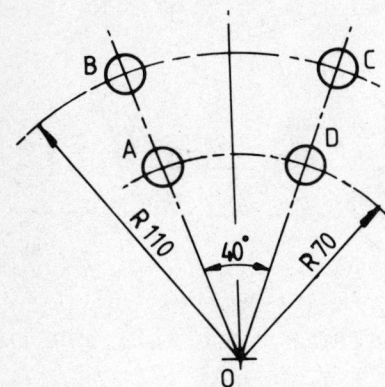

Figure 11.3

Applying the cosine rule to $\triangle$OAC

$$AC^2 = AO^2 + OC^2 - 2 \times AO \times OC \times \cos 40°$$
$$AC^2 = 70^2 + 110^2 - 2 \times 70 \times 110 \times \cos 40°$$
$$AC^2 = 4900 + 12\,100 - 15\,400 \cos 40°$$
$$AC^2 = 4900 + 12\,100 - 11\,800$$
$$AC^2 = 5200$$
$$AC = 72.1 \text{ mm}$$

The centre distance, AC = 72.1 mm

EXERCISE 11.1

1. Two forces on a vector diagram are represented by the sides AB = 297 N, CA = 106 N. The angle between the vectors $\angle$BAC = 40°. Calculate the magnitude of the resultant force BC.

2. The three sides of a triangle are 500 mm, 600 mm and 455 mm. Calculate the angle opposite the largest side.

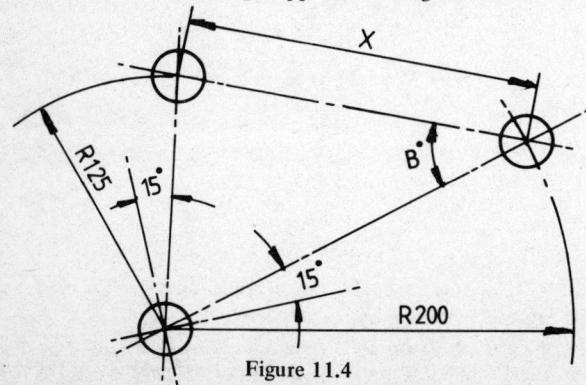

Figure 11.4

3. Figure 11.4 shows a jig-plate. Calculate the checking distance X between the centres of the two holes.

4. The three sides of a triangle are 3.5 m, 4.7 m and 6.8 m. Calculate the angle opposite the longest side.

11.2 AREA OF A TRIANGLE

The area of a triangle is equal to half the base times the perpendicular height.

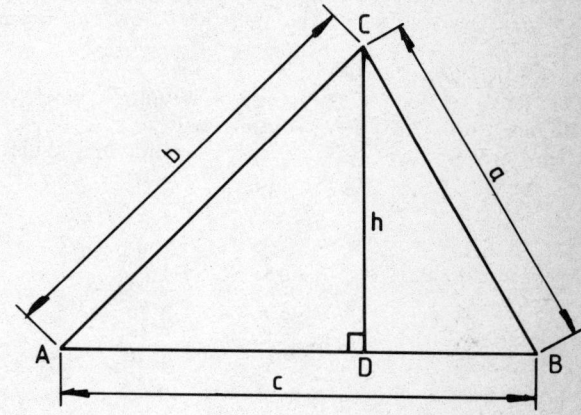

Figure 11.5

In $\triangle$ABC, fig. 11.5   area $= \frac{1}{2}ch$ (1)

But in $\triangle$ACD, $\angle$ADC = 90°

$$\sin A = \frac{h}{b} \quad \text{or} \quad h = b \sin A$$

substituting $h = b \sin A$ into (1) we have

$$\text{Area} = \frac{1}{2}cb \sin A$$

The area of a triangle is half the product of two sides times the sine of the angle between them.

Another formula for the area of a triangle, when the three sides are known is $\sqrt{s(s-a)(s-b)(s-c)}$ where $s = \dfrac{a+b+c}{2}$

Thus in any triangle the area can be found from:

Area $= \frac{1}{2}$(base) $\times$ (perpendicular height);

Area $= \frac{1}{2}bc \sin A$

Area $= \sqrt{s(s-a)(s-b)(s-c)}$ where $s = \dfrac{a+b+c}{2}$

Example 11.3 Find the area of the triangle ABC, given $a = 60$ mm, $b = 75$ mm and $C = 30°$.

$$\text{Area} = \frac{1}{2}ab \sin C$$
$$= \frac{1}{2} \times 60 \times 75 \times \sin 30°$$
$$= 1125 \text{ mm}^2$$

Example 11.4 Find the area of the triangle ABC, given $a = 1.4$ m, $b = 2.6$ m and $c = 3.1$ m.

$$s = \frac{1.4 + 2.6 + 3.1}{2} = 3.55$$

$$\text{Area} = \sqrt{3.55(3.55 - 1.4)(3.55 - 2.6)(3.55 - 3.1)}$$

$$= \sqrt{3.55 \times 2.15 \times 0.95 \times 0.45}$$

$$= \sqrt{3.263} = 1.806$$

$$\text{Area} = 1.806 \text{ m}^2$$

Example 11.5 In *Example 11.1* Find the depth of the point of suspension of the mass C below the beam AB.

In fig. 11.1, $\theta = 76°38'$

Let depth of point C below AB be h metres, which is the height of $\triangle$ABC if AB is the base.

Hence area of $\triangle$ABC $= \frac{1}{2} \times$ base $\times$ perpendicular height

$$= \frac{1}{2} \times 5.0 \times h$$

$$= 2.5h \tag{1}$$

Also area of $\triangle$ABC $= \frac{1}{2}ab \sin C$

$$= \frac{1}{2} \times 3.6 \times 4.4 \times \sin 76°38'$$

$$= 7.92 \sin 76°38'$$

$$= 7.705 \tag{2}$$

But the areas (1) and (2) are the same

Hence $2.5h = 7.705$

$$h = \frac{7.705}{2.5} = 3.08 \text{ m}$$

The point of suspension of the mass is 3.08 m below the beam.

EXERCISE 11.2

1. Find the areas of the following triangles:
(a) $a = 6.2$ m, $b = 3.9$ m and $C = 65°$
(b) $a = 120$ mm, $b = 80$ mm and $c = 70$ mm
(c) $c = 64$ mm, $b = 45$ mm and $A = 117°$
(d) $a = 1.25$ m, $b = 1.65$ m and $c = 2.35$ m

2. A triangle has sides of length 28 mm, 20 mm and 32 mm. If the longest side is the base of the triangle, calculate the perpendicular height.

3. Calculate the area of a cover plate which is in the shape of a quadrilateral ABCD. AB = 110 mm, BC = 154 mm, CD = 110 mm, DA = 132 mm and the diagonal AC = 176 mm.

4. A sheet of metal is a quadrilateral ABCD with sides AB = 0.80 m, BC = 0.89 m, CD = 0.58 m, DA = 0.49 m also $\angle$ABC = 49° and $\angle$CDA = 90°. Calculate the area of the sheet of metal. If the metal is 6 mm thick and has a density of $7.8 \times 10^3 \text{ kg/m}^3$ calculate the mass of metal in the sheet.

11.3 GENERAL EXAMPLES

Example 11.6 Five holes are drilled in a plate, as shown in fig. 11.6, all dimensions are in mm. Calculate the

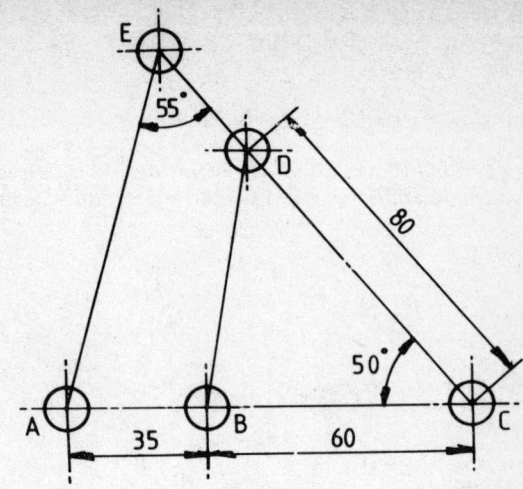

Figure 11.6

centre distances BD and DE. Consider $\triangle$BDC, using the cosine rule

$$BD^2 = BC^2 + DC^2 - 2 \times BC \times DC \times \cos \angle BCD$$

$$BD^2 = 60^2 + 80^2 - 2 \times 60 \times 80 \times \cos 50°$$

$$BD^2 = 3600 + 6400 - 9600 \times \cos 50°$$

$$BD^2 = 10\,000 - 6172 = 3828$$

$$BD = \sqrt{3828} = 61.87 \text{ mm}$$

By calculator

CM 60 $\times$ = M+ 80 $\times$ = M+ 50 Fcos $\times$ 2 $\times$ 80 $\times$ 60 = cs M+ RM $\sqrt{}$ 61.88

Consider $\triangle$AEC, applying the sine rule

$$\frac{EC}{\sin \angle EAC} = \frac{AC}{\sin \angle AEC}$$

But $\angle$EAC $= 180° - 50° - 55° = 75°$

$$\frac{EC}{\sin 75°} = \frac{95}{\sin 55°}$$

$$EC = \frac{95 \times \sin 75°}{\sin 55°} = 112.0 \text{ mm}$$

By calculator: CM 55 Fsin M+ 75 Fsin $\times$ 95 $\div$ RM = 112.0

But ED = EC $-$ DC = 112 $-$ 80 = 32 mm
Hence BD = 62 mm and ED = 32 mm

EXERCISE 11.3

1. Fig. 11.7 shows the positions of three holes to be drilled in a casting. Calculate the dimensions X and Y.

2. The belt drive on an internal combustion engine includes the engine pulley, fan pulley, and dynamo pulley, with centre

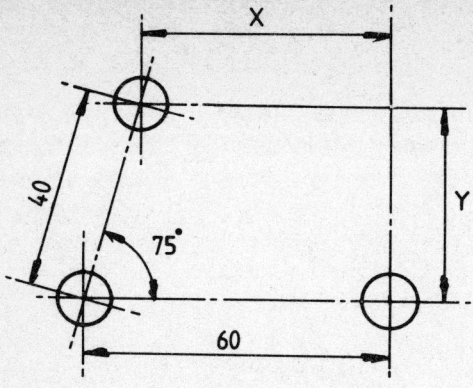

Figure 11.7

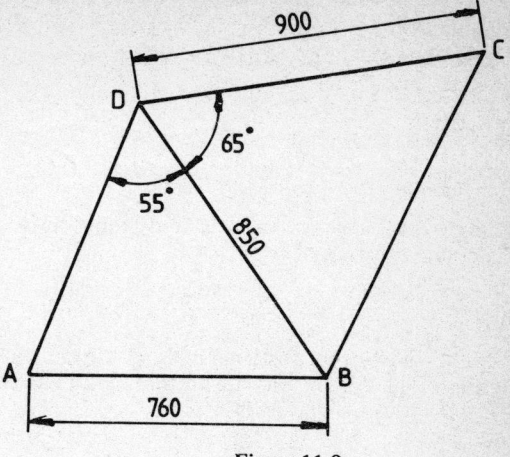

Figure 11.9

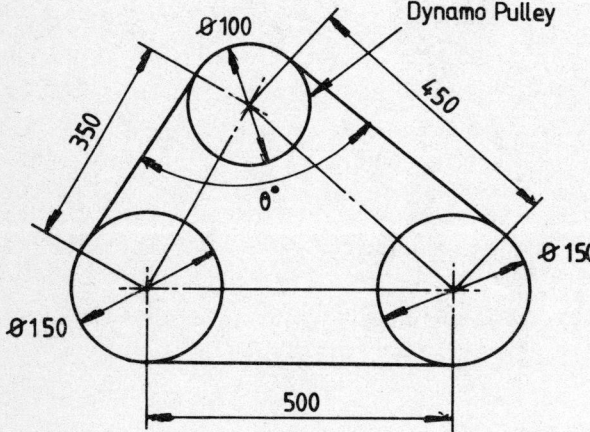

Dynamo Pulley

Figure 11.8

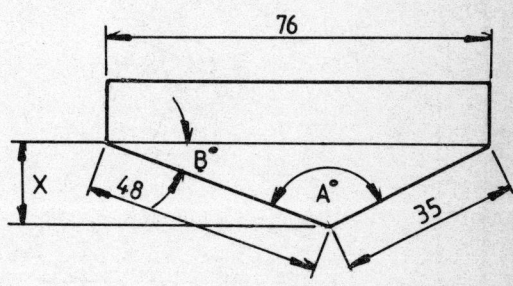

Figure 11.10

SUMMARY

1. The cosine rule states that in any triangle ABC
$$c^2 = a^2 + b^2 - 2ab \cos C.$$

2. The cosine rule is used to solve triangles when (a) 3 sides are given, to find the largest angle, then use the sine rule, (b) 2 sides and the included angle are given, to find the third side, then use the sine rule.

3. The area of a triangle ABC can be found as follows:
(a) given two sides and the included angle;
Area $= \frac{1}{2}ab \sin C$
(b) given 3 sides; area $= \sqrt{s(s-a)(s-b)(s-c)}$ where
$$s = \frac{a+b+c}{2}$$

SELF ASSESSMENT PAPER No 11

Instructions: Answer all questions in both sections
Time allowed: Section A 30 minutes
  (10 marks each question)
  Section B 30 minutes
  (20 marks each question)
Marks gained: 40+ pass with credit, 32–40 pass, less than 32 fail, repeat chapter 11.

distances as shown in fig. 11.8, all dimensions in mm. Find (a) the angle θ that the belt makes over the dynamo pulley, (b) the angle of lap the belt makes on the dynamo pulley.

3. A vertical aerial stands on ground rising uniformly at 6° to the horizontal. It is required to find the height of the aerial. To do this, two points A and B were selected 32 m apart on the sloping ground, and angles of elevation from the horizontal were found to be 42° and 52° respectively (the points A, B and the base of the aerial form a straight line). Calculate the height of the aerial.

4. Fig. 11.9 shows the outline of a frame structure. Determine the magnitude of the angle A, given it is acute, and the length of the member BC, all dimensions in mm.

5. Fig. 11.10 shows the plan view of part of a forming tool, all dimensions are in mm. Calculate the magnitudes of the angles A and B and the dimension X.

Section A

1. In triangle ABC, $a = 66$ mm, $b = 40$ mm and $c = 78$ mm. Calculate the largest angle of the triangle.

2. In triangle ABC, $a = 60$ mm, $b = 50$ mm and $C = 75°$. Calculate the length of side c.

3. In triangle ABC, $a = 90$ mm, $b = 70$ mm and $c = 120$ mm. Calculate the area of the triangle.

4. In triangle ABC, $a = 100$ mm, $b = 80$ mm and $C = 105°$. Calculate the area of the triangle.

Section B

1. Fig. 11.11 shows the position of four holes drilled in a template, all dimensions are in mm. Find (a) the length of BD, (b) the angle BCD.

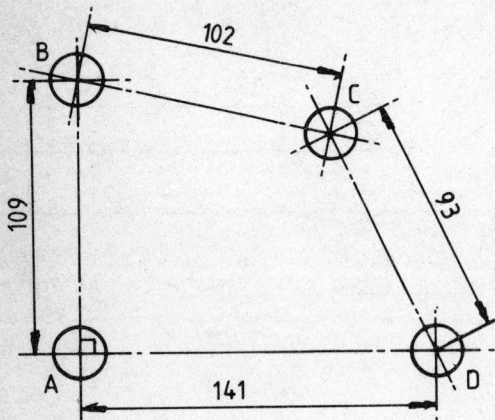

Figure 11.11

2. A weight hangs from the junction of two ropes 2.2 m and 1.8 mm long respectively. If the other ends of the ropes are attached to a beam, as shown in fig. 11.12, calculate the depth of C, the junction of the ropes, below the beam AB. Also find the angle ACB.

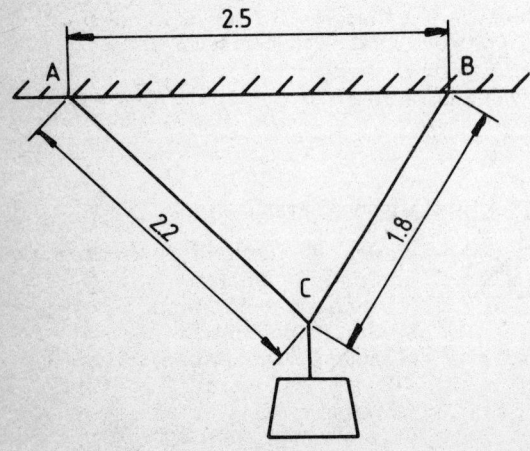

Figure 11.12

ANSWERS

Exercise 11.1   1.   226 N,   2. $77°38'$,
3.   175 mm,   4. $111°12'$

Exercise 11.2   1.(a) 10.96 m^2,   (b 2690 mm^2,
(c) 1283 mm^2,   (d) 0.984 m^2,   2.   17.32 mm,
3. $15\,634 \text{ mm}^2$,   4. 0.411 m^2, 19.2 kg.

Exercise 11.3   1. $x = 49.7$ mm, $y = 38.6$ mm,
2.   (a) $83°29'$,   (b) $96°31'$,   3.   78 m,
4. $66°22'$, 941 mm,   5. $A = 132°3'$,
$B = 20°1'$, $x = 16.4$ mm

SELF ASSESSMENT PAPER No 11

Marks

Section A

1. Largest angle opposite side 78 mm   2

$$\cos C = \frac{66^2 + 40^2 - 78^2}{2 \times 66 \times 40}$$ 4

$\cos C = -0.0243$ 2
$C = 91°23'$ 2

2. $c^2 = 60^2 + 50^2 - 2 \times 60 \times 50 \cos 75°$ 4
$c^2 = 6100 - 6000 \cos 75°$ 2
$c^2 = 4547$ 2
$c = 67.4$ mm   2

3. $s = \dfrac{90 + 70 + 120}{2} = 140$ mm   2

Area $= \sqrt{140(140 - 90)(140 - 70)(140 - 120)}$ 4
$=$
$= \sqrt{140 \times 50 \times 70 \times 20}$ 2
$= 3130 \text{ mm}^2$ 2

4. Area $= \frac{1}{2} \times 100 \times 80 \times \sin 105°$ 6
$= 4000 \sin 105°$ 2
$= 3864 \text{ mm}^2$ 2

Section B

1. (a) In $\triangle$ABD, $\angle$BAD $= 90°$ 2
$BD^2 = 109^2 + 141^2$ 4
$BD = 178$ mm   2

(b) In $\triangle$BCD, $\cos C = \dfrac{102^2 + 93^2 - 178^2}{2 \times 102 \times 93}$ 6

$\cos C = -0.6658$ 4
$\angle$BCD $= 131°44'$ 2

2. Area of $\triangle$ACB $= \frac{1}{2} \times 2.5 \times h$, where $h =$ depth of C below AB   2

In $\triangle$ACB, $s = \dfrac{2.5 + 1.8 + 2.2}{2} = 3.25$ 1

Area of $\triangle$ACB $=$
$= \sqrt{3.25(3.25 - 2.5)(3.25 - 1.8)(3.25 - 2.2)}$ 3

$= \sqrt{3.25 \times 0.75 \times 0.45 \times 1.05}$ 1

$= 1.073$ 1

$\frac{1}{2} \times 2.5 \times h = 1.073$ 2

$h = 0.859$ m 2

$\cos C = \dfrac{2.2^2 + 1.8^2 - 2.5^2}{2 \times 2.2 \times 1.8}$ 4

$\cos C = 0.2311$ 2

$\angle ACB = 76°38'$ 2

12 Trigonometry, Practical Applications

12.1 MEASUREMENT OF TAPERS

For external tapers two sets of similar rollers are held in contact with the taper on opposite sides as shown in fig. 12.1. Formulae can be derived for these measurements but they are complicated to remember. It is better to draw a clear diagram and calculate each example separately.

Example 12.1 A taper is 80 mm long and rollers of 20 mm diameter are used to check the taper. The dimensions are as given in fig. 12.1, all dimensions are in mm. Calculate (a) the taper angle, (b) the diameter of the taper 30 mm from the smaller end.

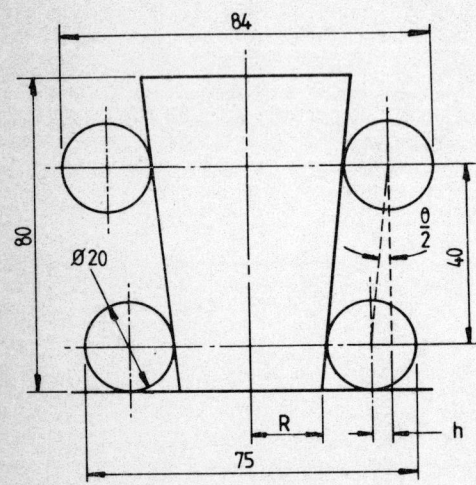

Figure 12.1

(a) h is the horizontal distance between the centres of the rollers on one side. Hence $h = \dfrac{84 - 75}{2} = 4.5$ mm

$$\tan \frac{\theta}{2} = \frac{h}{40} = \frac{4.5}{40} = 0.1125$$

$$\frac{\theta}{2} = 6.419° \ (6°25')$$

$$\theta = 12°50'$$

The taper angle is $12°50'$

(The taper can also be expressed as 9 mm in 40 mm or 1 in 4.44.)

(b) In fig. 12.2 (a) $R = PQ = PT - ST - RS - QR$

$$PT = \frac{75}{2} = 37.5 \text{ mm}, \ ST = 10 \text{ mm},$$

$$RS = \frac{10}{\cos 6°25'}, \ QR = 10 \tan 6°25'$$

Hence $R = 37.5 - 10 - \dfrac{10}{\cos 6°25'} - 10 \tan 6°25'$

$$R = 37.5 - 10 - 10.06 - 1.12$$

$$R = 16.32$$

In fig. 12.2(b) $r = R + AB = 16.32 + 30 \tan 6°25'$

$$r = 16.32 + 3.37 = 19.69 \text{ mm}$$

The diameter of the taper 30 mm from the smaller end is 39.38 mm.

Internal tapers can be checked by the two ball method. Again it is better to calculate each time rather than try to use a formula.

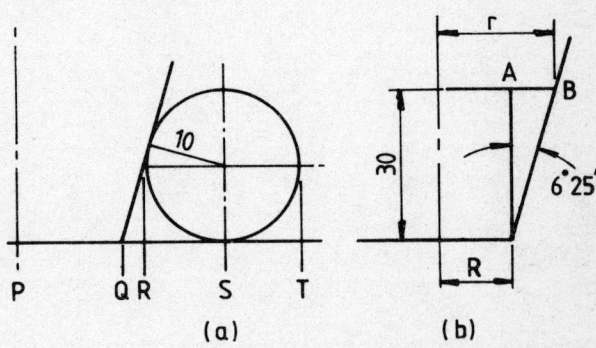

Figure 12.2

Example 12.2 The included angle of an internal taper component is measured using the two ball method, as shown in fig 12.3(a), all dimensions are in mm. Working from first principles determine the value of the included angle of the taper.

$$AB = 280 - 37.5 - 64 + 25 = 203.5 \text{ mm}$$

In $\triangle EBA$, $AB = 203.5$ mm

$$EA = AD - BC = 37.5 - 25 = 12.5 \text{ mm}$$

since $\angle AEB = 90°$, $\sin \angle EBA = \dfrac{EA}{AB} = \dfrac{12.5}{203.5} = 0.0614$

Hence $\angle EBA = 3.52° \ (3°31')$

Included angle of the taper $= 2 \times \angle EBA = 7.04° \ (7°2')$

The taper is $7°2'$

(The taper can also be expressed as 25 mm in 203.5 mm or 1 mm in 8.1 mm.)

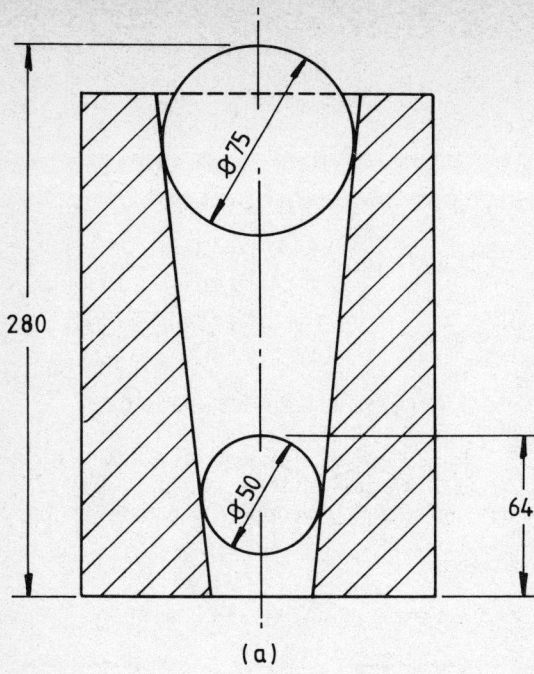

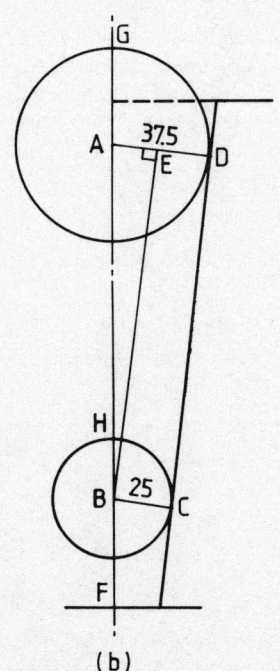

(a)

(b)

Figure 12.3

In fig. 12.3(b), the included angle of the taper is $2 \times \angle EBA$

$$AB = FG - AG - HF + HB$$

EXERCISE 12.1

1. For the taper plug, shown in fig. 12.4, all dimensions being in mm, calculate (a) the taper angle, (b) the top diameter of the plug. The rollers are 24 mm diameter.

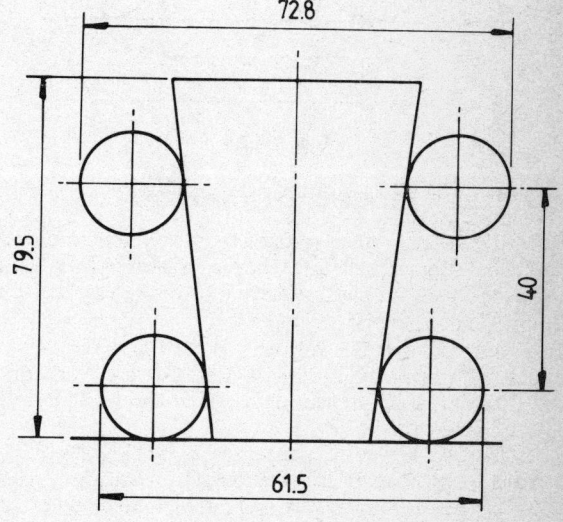

Figure 12.4

2. Rollers of 20 mm diameter are used to check a taper plug with a 12° taper, as shown in fig. 12.5, all dimensions in mm. Calculate the checking dimensions X and Y.

3. Two balls of diameter 25 mm and 40 mm are used to check an internal taper. If the distance between the ball centres is 55 mm calculate the taper angle.

4. Two balls of diameters 30 mm and 50 mm are used to check a taper hole. The depth of the hole is 200 mm, the distance from the top of the hole to the top of the 30 mm ball is 120 mm and the distance from the top of the hole to the top of the 50 mm ball is 10 mm. Calculate the angle of taper and also express the taper in the form 1 in k. Calculate the diameter of the bottom of the hole.

5. A tapered hole in a metal block which has been machined on all faces has a maximum diameter of 34.3 mm and an included angle of 22°. A precision ground ball, 22.2 mm diameter, is placed in the hole. Calculate: (a) the distance between the top surface of the block and the highest point of the ball, (b) the taper on the diameter of the hole in millimetres per metre length.

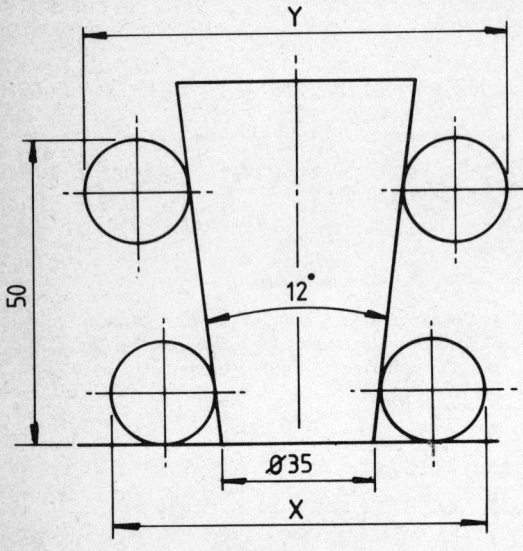

Figure 12.5

12.2 OTHER PRACTICAL EXAMPLES

Example 12.3 A 90° vee-gauge is placed over a circular bar. The shortest distance between the corner of the vee-gauge and the circumference of the bar is 20.7 mm. Find the diameter of the bar.

The shortest distance will be AB, in fig. 12.6, this distance is given as 20.7 mm. OC and OD are the radii of the circular section joining the centre to the points of contact of the bar.

$\angle ACO = \angle ADO = 90°$, and since $\angle CAD = 90°$ the quadrilateral CADO is a rectangle. Also since CO = OD = r the rectangle is a square of side r, where r is the radius of the section.

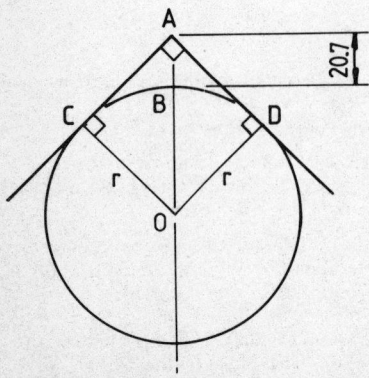

Figure 12.6

In $\triangle ADO$, $\angle ADO = 90°$ $\sin 45° = \dfrac{OD}{OA}$

$$\sin 45° = \frac{r}{20.7 + r}$$

$$\text{or } (20.7 + r) \sin 45° = r$$

$$(20.7 + r)0.7071 = r$$

$$14.64 + 0.7071r = r$$

$$r - 0.7071r = 14.64$$

$$0.2929r = 14.64$$

$$r = 50.0 \text{ mm}$$

The radius of the bar is 50.0 mm

Example 12.4 An arrangement for checking the dimensions of a dovetail slide is shown in fig. 12.7. The rollers are 50 mm in diameter. Calculate the dimension X.

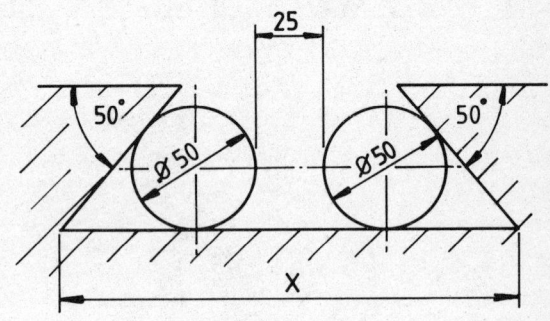

Figure 12.7

Consider one roller as in fig. 12.8.
In $\triangle AOB$, $\angle ABO = 90°$ and since $\angle DAB = 50°$, $\angle OAB = 25°$

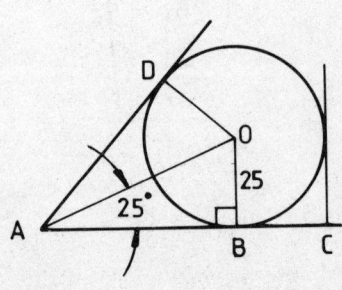

Figure 12.8

$$\tan 25° = \frac{OB}{AB} = \frac{25}{AB}$$

$$AB = \frac{25}{\tan 25°} = \frac{25}{0.4663} = 53.61 \text{ mm}$$

$$X = 25 + 2 \times AC = 25 + 2 \times (AB + BC)$$

$$X = 25 + 2 \times (53.6 + 25)$$

$$X = 182.2 \text{ mm}$$

The dimension X is 182 mm

Example 12.5 Determine the vectorial sum of the currents given by $I_1 = 20 \sin\theta$, $I_2 = 30 \sin\left(\theta + \frac{\pi}{3}\right)$

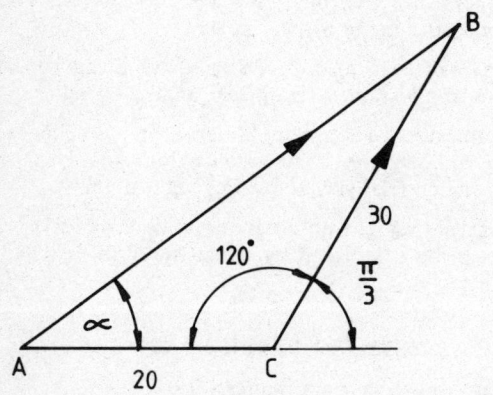

Figure 12.9

The currents can be represented as shown in fig. 12.9, the sum being represented by the vector AB.
The magnitude of AB can be found by considering $\triangle$ABC with AC = 20, CB = 30 and $\angle$ACB = 120°
Using the cosine rule

$$AB^2 = AC^2 + CB^2 - 2 \times AC \times CB \times \cos 120°$$

$$AB^2 = 20^2 + 30^2 - 2 \times 20 \times 30 \times -0.5$$

$$AB^2 = 400 + 900 + 600 = 1900$$

$$AB = \sqrt{1900} = 43.6$$

Using the sine rule $\dfrac{43.6}{\sin 120°} = \dfrac{30}{\sin \alpha}$

$$\sin \alpha = \frac{30 \times \sin 120°}{43.6} = 0.5959$$

$$\alpha = 36.58° \text{ (0.638 radians)}$$

Hence AB is $43.6 \sin(\theta + 0.638)$

The sum of the two currents I_1 and I_2 is $43.6 \sin(\theta + 0.638)$

EXERCISE 12.2

1. A 120° vee-gauge rests on a circular bar and the corner of the vee is found to be 11.9 mm from the circumference of the bar. Calculate the diameter of the bar.

2. A 100° vee-gauge is used to find the diameter of a cylinder. The distance between the corner of the vee and the cylinder is found to be 18.64 mm, calculate the diameter of the cylinder.

3. Two 30 mm rollers are used to check the dovetail slide, shown in fig. 12.10, all dimensions are in mm. Calculate the dimension X.

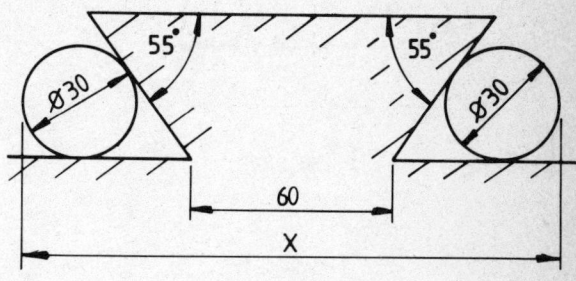

Figure 12.10

4. A roller, 100 mm diameter, is placed in a symmetrical 90° vee-block, see fig. 12.11, all dimensions in mm. If the vee is 280 mm wide at the top calculate the dimension X.

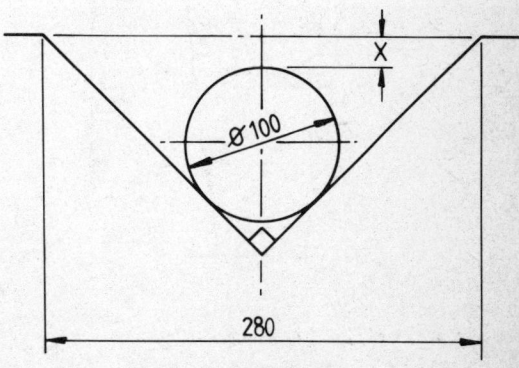

Figure 12.11

5. Find the vectorial sum of the currents $I_1 = 30 \sin\theta$ and $I_2 = 50 \sin\left(\theta + \frac{\pi}{4}\right)$.

6. Determine the vectorial sum of I_1 and I_2 when $I_1 = 25 \sin\theta$ and $I_2 = 40 \sin\left(\theta + \frac{\pi}{6}\right)$.

12.3 GENERAL EXAMPLES

EXERCISE 12.3

1. A taper plug is 120 mm long and is checked with rollers, as shown in fig. 12.12, all dimensions in mm. The rollers are of radius 20 mm, calculate the taper angle and the diameter of the top of the plug.

2. Two balls of diameters 30 mm and 20 mm are used to test a taper hole, as shown in fig. 12.13, all dimensions in mm. Calculate the angle of the taper and the dimension x.

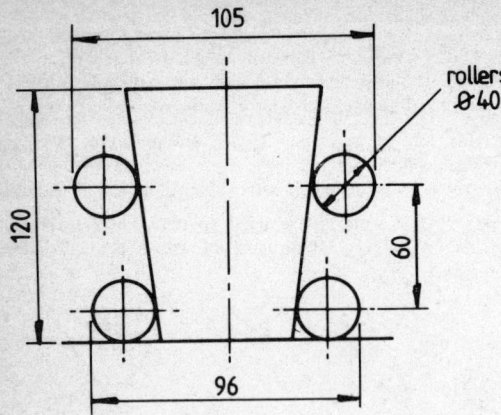

Figure 12.12

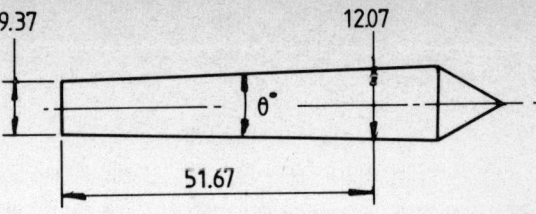

Figure 12.15

4. Find the vectorial sum of the currents $I_1 = 15 \sin (\theta + \frac{\pi}{6})$ and $I_2 = 30 \sin (\theta + \frac{\pi}{4})$.

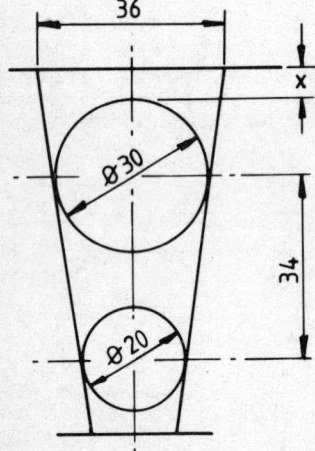

Figure 12.13

3. (a) Two 12 mm diameter rollers are used to check the dovetail slide as shown in fig. 12.14. Calculate d. (b) A morse tapered lathe centre is to be made to the dimensions shown in fig. 12.15. Find, by calculation, the included angle θ, all dimensions in mm.

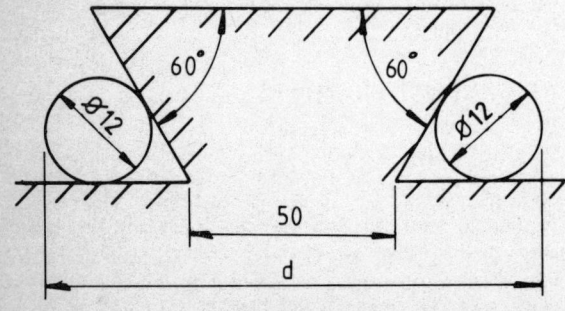

Figure 12.14

SUMMARY

1. Tapers can be calculated by forming right angled triangles and using trigonometry.

2. Dovetails can also be checked by using rollers and calculating distances from right angled triangles.

3. Diameters of large cylinders can be checked by using a vee-gauge. The calculations involve forming right angled triangles and using trigonometry.

4. Alternating currents can be added vectorially. The sum can be calculated by using the cosine rule and the sine rule.

SELF ASSESSMENT PAPER No 12

Instructions: Answer all questions
Time allowed: One hour (20 marks each question)
Marks gained: 30+ pass with credit, 24—30 pass, less than 24 fail, repeat chapter 12.

1. A tapered hole is checked by using two balls of diameters 20 mm and 10 mm. The distance from the top of the balls to the top of the hole are 5.00 mm and 41.96 mm. Determine the taper angle of the hole.

2. The diameter of a cylinder is checked with a $120°$ vee-gauge. The distance between the corner of the vee-gauge and the cylinder is measured as 19.76 mm. Calculate the diameter of the cylinder.

3. Calculate the sum of the two currents $I_1 = 50 \sin \theta$ and $I_2 = 70 \sin \left(\theta + \frac{\pi}{6} \right)$.

ANSWERS

Exercise 12.1   1.   (a) $16°14'$, (b) 32.4 mm.
2.   77.2 mm, 83.6 mm,   3. $15°40'$,
4. $11°29'$, 1 in 5, 17.1 mm,   5.   (a) 18.9 mm,
(b) 194 mm/m

Exercise 12.2   1.   154 mm,   2.   122 mm,
3.   148 mm,   4.   19 mm,
5.   74.3 sin $(\theta + 0.496)$,   6.   62.9 sin $(\theta + 0.323)$

Exercise 12.3 1. $8°35'$, 30.9 mm, 2. $16.91°$, 19.1 mm, 3. (a) 82.8 mm, (b) $3°$, 4. $44.66 \sin(\theta + 0.7)$

SELF ASSESSMENT PAPER No 12

Marks

1. Distance between ball centres
$= 41.96 - 5.00 + 5.00 - 10.00$ 6
$= 31.96$ mm   1

$\tan \dfrac{\theta}{2} = \dfrac{10 - 5}{31.96}$ 6

$\tan \dfrac{\theta}{2} = 0.1564$ 2

$\dfrac{\theta}{2} = 8.89°$ 2

$\theta = 17.78°$ $(17°47')$ 3

2. $\sin 60° = \dfrac{r}{r + 19.76}$ 6

$(r + 19.76)\sin 60° = r$ 3
$r(1 - \sin 60°) = 19.76 \sin 60°$ 3

Marks

$r = \dfrac{19.76 \sin 60°}{1 - \sin 60°}$ 2

CM 60 Fsin M+ cs + 1 = (X ↔ M) × 19.76 ÷ RM
= 127.7
$r = 128$ mm   6

3. Sum $= I \sin(\theta + \alpha)$ 1

$\dfrac{\pi}{6} = 30°$, $180° - 30° = 150°$ 1

$I = \sqrt{50^2 + 70^2 - 2 \times 50 \times 70 \times \cos 150°}$ 5
CM 50 x = M+ 70 x = M+ 150 Fcos × 2 × 50 × 70 x =
cs M+ RM $\sqrt{}$ = 116
$I = 116$ 2

$\dfrac{70}{\sin \alpha} = \dfrac{116}{\sin 150°}$ 3

$\sin \alpha = \dfrac{70 \times \sin 150°}{116}$ 2

$\alpha = 17.56°$ 3
$\alpha = 0.307$ radians   2
Sum is $116 \sin(\theta + 0.307)$ 1

13 Three Dimensional Triangulation Problems

13.1 ANGLE BETWEEN A LINE AND A PLANE

The line AB, in fig. 13.1, meets the plane PQRS at the point A. The perpendicular from B to the plane PQRS meets the plane at C. The angle BAC is defined as the angle between the line AB and the plane PQRS.

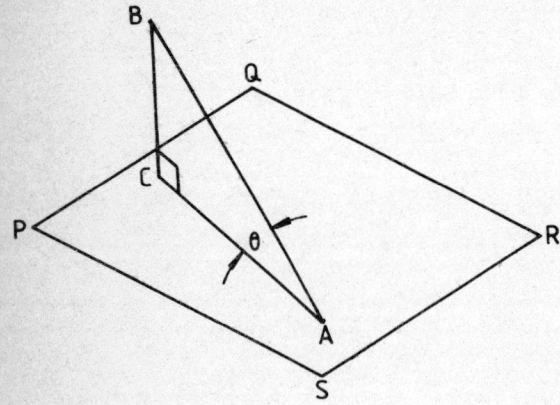

Figure 13.1

The point C is called the projection of the point B on the plane, and the line AC is the projection of AB on the plane.

Example 13.1 A rectangular metal block has dimensions 24.6 mm by 31.8 mm by 22.9 mm. Calculate the length of AC, the length of AP and the angle PAC in fig. 13.2.

$\triangle ABC$ has a right angle at B, using Pythagoras' Theorem

$$AC^2 = AB^2 + BC^2 = 24.6^2 + 31.8^2 = 1616.4$$

$$AC = \sqrt{1616.4} = 40.2 \text{ mm}$$

$\triangle ACP$ is right angled at C, using the theorem of Pythagoras

$$AP^2 = AC^2 + CP^2 = 40.2^2 + 22.9^2 = 2140.5$$

$$AP = \sqrt{2140.5} = 46.3 \text{ mm}$$

Also in $\triangle ACP$, $\tan \theta = \dfrac{CP}{AC} = \dfrac{22.9}{40.2} = 0.5697$

$$\theta = 29.67° \ (29°40')$$

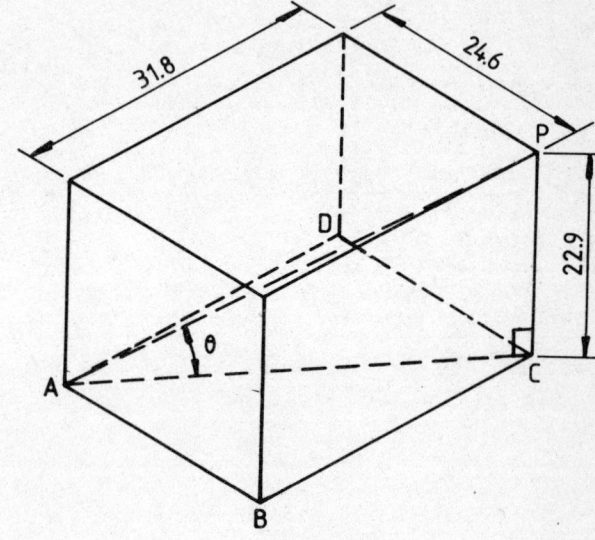

Figure 13.2

Hence AC = 40.2 mm, AP = 46.3 mm, $\angle PAC = 29°40'$
In fig. 13.1, AC is the plan length of AB.
$AC = AB \cos \theta$ and since $\cos \theta$ is always less than 1, the plan length must always be less than the true length. Similarly if we know the plan length AC the true length can be found from $AB = \dfrac{AC}{\cos \theta}$

Example 13.2 A rectangular sheet of metal 1.0 m by 0.85 m is inclined at an angle of 40°, as shown in fig. 13.3.
Calculate the inclination of the diagonal, AC, to the horizontal.
P is the foot of the perpendicular from C to the horizontal plane ABP.
Then AP is the projection of AC on this horizontal plane.
The required angle is $\angle CAP$.

From $\triangle ABC$, since $\angle ABC = 90°$, $AC^2 = AB^2 + BC^2$

$$AC^2 = 1.0^2 + 0.85^2 = 1.7225$$

$$AC = \sqrt{1.7225} = 1.31 \text{ m}$$

From △BPC, since ∠BPC = 90°, $\sin 40° = \dfrac{CP}{BC} = \dfrac{CP}{0.85}$

$$CP = 0.85 \times \sin 40° = 0.546 \text{ m}$$

From △APC, since ∠APC = 90°, $\sin \theta = \dfrac{CP}{AC} = \dfrac{0.546}{1.31}$

$$= 0.4168$$

$$\theta = 24.63° \ (24°38')$$

The diagonal AC is inclined at an angle of 24°38′ to the horizontal.

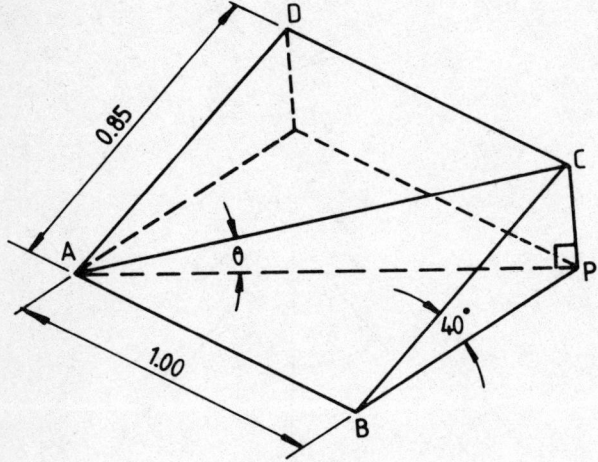

Figure 13.3

EXERCISE 13.1

1. The edges of a rectangular box are 30 mm, 60 mm and 70 mm. Calculate the values of the three angles that a diagonal makes with the faces of the box.

2. A square sheet of metal of side 1.24 m has one edge horizontal. The sheet of metal slopes at 35° to the horizontal. Find the angle between a diagonal of the square and the horizontal.

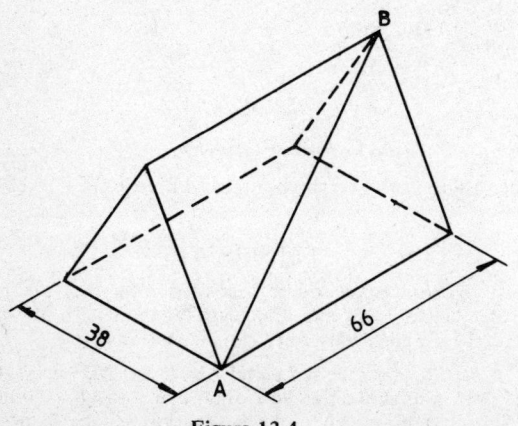

Figure 13.4

3. The square base of a square pyramid has sides 80 mm long. If each of the sloping edges has length 120 mm, calculate the perpendicular height of the pyramid.

4. A triangular prism has an equilateral triangle as its cross-section. It has the dimensions shown in fig. 13.4, all dimensions in mm. Calculate the length of the diagonal AB.

13.2 THE ANGLE BETWEEN TWO PLANES

Two planes which are not parallel intersect in a straight line. If two lines, one in each plane and each perpendicular to the line of intersection, are drawn, the angle between these lines is defined as the angle between the planes.

In fig. 13.5, the planes ABCD and ABXY intersect along AB. OP is in plane ABCD and is perpendicular to AB.

OQ is in plane ABXY and is perpendicular to AB. $\theta = \angle POQ = $ angle between the planes

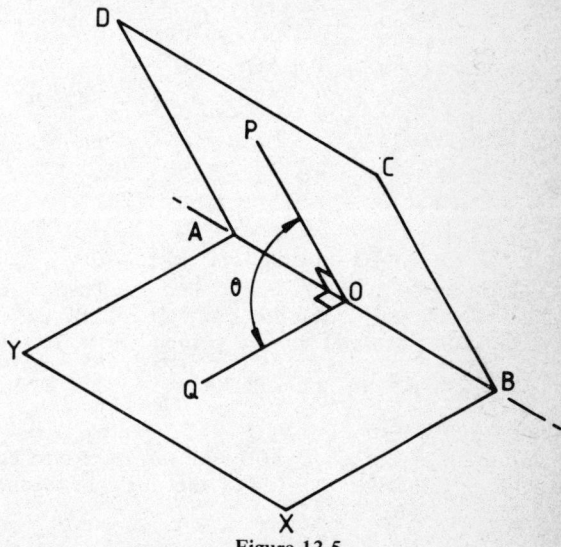

Figure 13.5

If plane ABXY is horizontal then OP is sometimes referred to as the line of greatest slope. All other lines from O in the plane ABCD will make a smaller angle with the plane ABXY. OP is the steepest line that can be drawn through O in the plane ABCD.

Example 13.3 A pyramid has a rectangular base 60 mm by 50 mm and a perpendicular height of 70 mm. Calculate the lenght of one of the sloping edges. Also find the angle this edge makes with the base and the angle between a sloping face and the base.

In △ABC, ∠ ABC = 90°

$$AC^2 = AB^2 + BC^2 = 50^2 + 60^2 = 6100$$

$$AC = \sqrt{6100} = 78.1 \text{ mm}$$

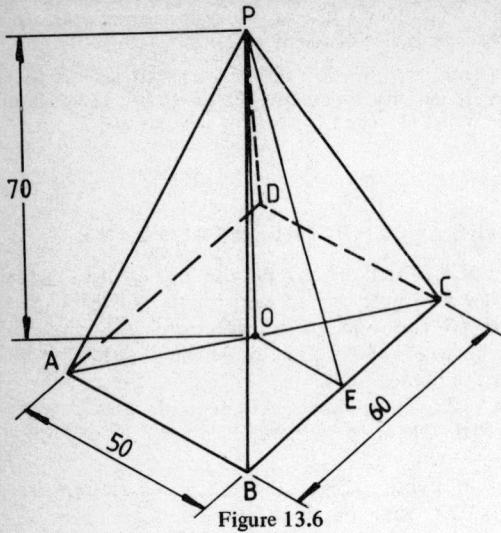

Figure 13.6

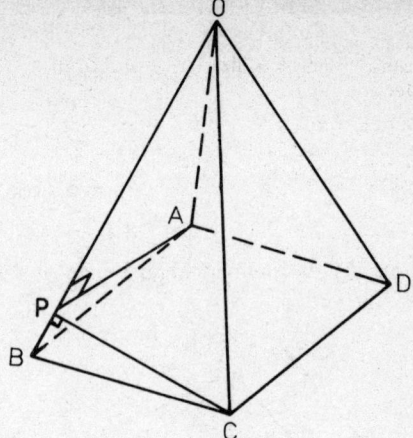

Figure 13.7

$$AO = \tfrac{1}{2}AC = 39.05 \text{ mm}$$

In $\triangle AOP$, $\angle AOP = 90°$

$$AP^2 = AO^2 + OP^2 = 39.05^2 + 70^2 = 6424.9$$

$$AP = \sqrt{6424.9} = 80.2 \text{ mm}$$

$$\tan \angle PAO = \frac{OP}{AO} = \frac{70}{39.05} = 1.7722$$

$$\angle PAO = 60.56° \ (60°34')$$

The angle between the face BPC and the base is the angle PEO where E is the mid point of BC, PE is perpendicular to BC and EO is perpendicular to BC.

In $\triangle OEP$, $\angle POE = 90°$, $\tan \angle PEO = \dfrac{OP}{OE} = \dfrac{70}{25} = 2.8$;

$\angle PEO = 70.35° \ (70°21')$

A sloping edge is of length 80.2 mm, the angle between an edge and the base is $60°34'$, the angle between a face and the base is $70°21'$.

Example 13.4 OABCD is a square pyramid with apex O and square base ABCD of side 80 mm. If each of the sloping edges is also 80 mm in length find the angle between the planes OAB and OBC.

The angle between the planes OAB is the angle APC, shown in fig. 13.7. Where PC is perpendicular to OB and PA is perpendicular to OB.

Since OB = BC = OC, $\triangle OBC$ is equilateral.

Hence $\angle OBC = 60°$, and P will be the mid point of OB.

In $\triangle PBC$, $PC^2 = BC^2 - PB^2 = 80^2 - 40^2 = 4800$

$$PC = \sqrt{4800} = 69.3 \text{ mm}$$

Similarly AP = 69.3 mm

In $\triangle ABC$, $\angle ABC = 90°$, $AC^2 = AB^2 + BC^2$

$$= 80^2 + 80^2 = 12\,800$$

$$AC = \sqrt{12\,800} = 113.1 \text{ mm}$$

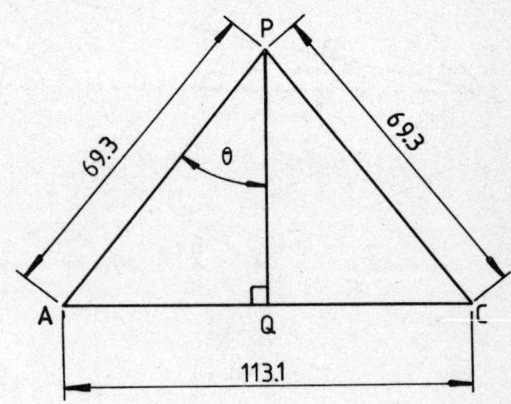

Figure 13.8

In $\triangle APC$, AP = PC = 69.3 mm and AC = 113.1 mm

In fig. 13.8, $AQ = \tfrac{1}{2}AC = 56.55$ mm

In $\triangle APQ$, $\angle AQP = 90°$

$$\sin \theta = \frac{AQ}{AP} = \frac{56.55}{69.3} = 0.8160$$

$$\theta = 54.69° \ (54°41')$$

$$\angle APC = 2\theta = 109.38° \ (109°22')$$

The angle between the planes OAB and OBC is $109°22'$.

EXERCISE 13.2

1. A pyramid has a square base of side 64 mm. The perpendicular height is 80 mm. Calculate (a) the length of a sloping edge, (b) the angle between a sloping face and the base.

2. A rectangular sheet of metal ABCD has AB horizontal and CD is 15 mm above this horizontal plane. If AB = 80 mm and BC = 40 mm calculate the inclination of the plane to the horizontal.

3. A frustum of a square pyramid has a perpendicular height 45 mm. The square base has side 50 mm and the square top has side 30 mm. Calculate (a) the length of a sloping edge, (b) the angle between a sloping edge and the base, (c) the angle between a sloping face and the base.

4. Fig. 13.9 shows a double wedge with CD perpendicular to the base ABC. AC = BC = 50 mm, CD = 30 mm and ∠ACB = 55°. Calculate the angle between the plane ABD and the base ABC.

5. OABCD is a square pyramid and the square base has sides of length 60 mm. If each of the sloping edges is also of length 60 mm calculate (a) the angle between a sloping edge and the base, (b) the angle between a sloping face and the base, (c) the angle between the sloping faces OAB and OBC, (d) the angle between the sloping faces OAB and OCD.

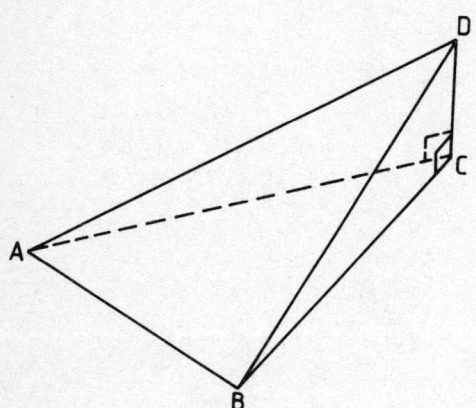

Figure 13.9

13.3 GENERAL EXAMPLES

Example 13.5 A vertical mast AB stands on horizontal ground. From a point P, 20 m above the ground, the angle of elevation of A is 55° and the angle of depression of B is 22°, see fig. 13.10. Calculate

(a) the distance BQ, (b) the height of the mast AB, (c) the angle of elevation of A from Q.

The angle of elevation of A from P is the angle between the horizontal and the line AP. The angle of depression of B from P is the angle between the horizontal and the line PB.

(a) In △PQB, ∠PQB = 90°, ∠PBQ = 22°

$$\tan \angle PBQ = \frac{PQ}{QB}, \quad \tan 22° = \frac{20}{QB}$$

$$QB = \frac{20}{\tan 22°} = 49.5 \text{ m}$$

(b) In △APC, PC = QB = 49.5 m, ∠ACP = 90°

$$\tan 55° = \frac{AC}{PC}, \quad \tan 55° = \frac{AC}{49.5}$$

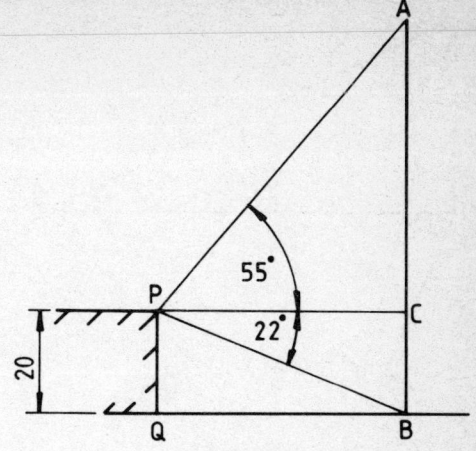

Figure 13.10

$$AC = 49.5 \times \tan 55° = 70.7 \text{ m}$$

In △PCB, $\tan 22° = \dfrac{BC}{PC}$, $\tan 22° = \dfrac{BC}{49.5}$

$$BC = 49.5 \times \tan 22° = 20.0 \text{ m}$$

Height of mast = AC + BC = 70.7 + 20.0 = 90.7 m

(c) The angle of elevation of A from Q is the angle AQB.

$$\tan \angle AQB = \frac{AB}{QB} = \frac{90.7}{49.5} = 1.832$$

$$\angle AQB = 61.38° \; (61°23')$$

The angle of elevation of A from Q is 61°23′

Example 13.6 From a point A due south of a tower, the angle of elevation of the top of the tower is 6°30′, see fig. 13.11. From a point B due east of the tower, the angle of elevation of the top of the tower is 9°45′. If the distance AB = 200 m calculate the height of the tower.

Let the height of the tower be *h* metres.

In △POA, ∠POA = 90°, $\tan 6°30' = \dfrac{OP}{OA}$

$$\tan 6°30' = \frac{h}{OA}, \quad OA = \frac{h}{\tan 6°30'} = \frac{h}{0.1139}$$

$$OA = 8.777h$$

Similarly in △POB, ∠POB = 90°

$$\tan 9°45' = \frac{OP}{OB} = \frac{h}{OB}, \quad OB = \frac{h}{\tan 9°45'} = \frac{h}{0.1718}$$

$$OB = 5.820h$$

In △OAB, ∠AOB = 90°, Using Pythagoras

$$AB^2 = OA^2 + OB^2$$

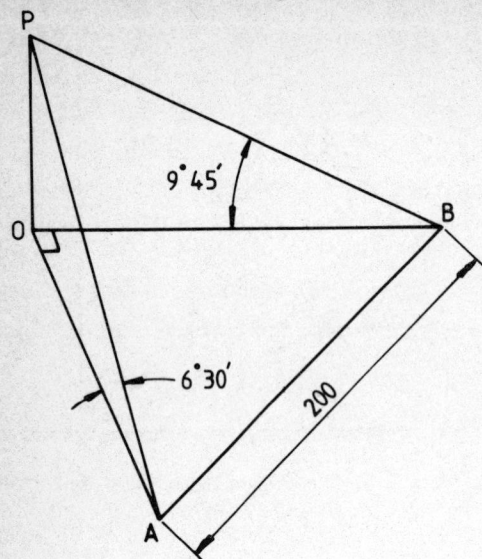

Figure 13.11

but AB = 200, OA = 8.777h, OB = 5.820h

$$200^2 = (8.777h)^2 + (5.820h)^2$$

$$40\,000 = 77.04h^2 + 33.87h^2$$

$$40\,000 = 110.91h^2$$

$$h^2 = \frac{40\,000}{110.91} = 360.64$$

$$h = \sqrt{360\,64} = 19 \text{ m}$$

The height of the tower is 19 m

Example 13.7 ABCD$_1$ is a rectangle in which AB = 80 mm and BC = 60 mm. E is the foot of the perpendicular from B to AC.
Calculate the lengths of BE and D$_1$E, see fig. 13.12. If the rectangle is folded about AC so that the plane ACD$_2$ is perpendicular to the plane ABC, calculate the length of BD$_2$, and the angle BD$_2$E.

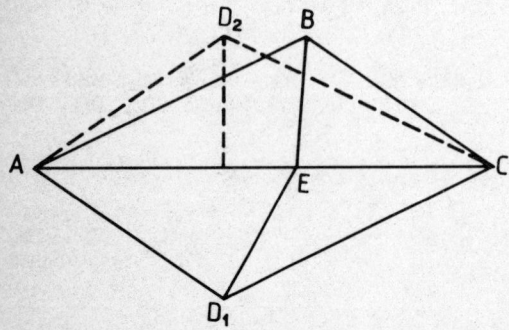

Figure 13.12

In $\triangle ABC$, $\angle ABC = 90°$, AB = 80 mm, BC = 60 mm and hence AC = 100 mm.

$$\sin \angle BAC = \frac{60}{100} = 0.6, \angle BAC = 36.87°$$

In $\triangle ABE$, $\angle BAE = 36.87°$, $\angle BEA = 90°$

$$\sin \angle BAE = \frac{BE}{AB} \text{ or } BE = AB \sin \angle BAE$$

$$BE = 80 \times \sin 36.87° = 48 \text{ mm}$$

In $\triangle ACD_1$, $\angle AD_1C = 90°$, $\cos \angle CAD_1 = \frac{D_1C}{AD_1} = \frac{60}{100}$

$$= 0.6 \quad \angle CAD_1 = 53.13°$$

In $\triangle ABE$, $\angle BEA = 90°$, $AB^2 = AE^2 + EB^2$
or $AE^2 = AB^2 - EB^2$

$$AE^2 = 80^2 - 48^2 = 4096$$

$$AE = \sqrt{4096} = 64 \text{ mm}$$

In $\triangle AED_1$, $\angle D_1AE = 53.13°$, $AE = 64$, $AD_1 = 60$

Using the cosine rule $ED_1^2 = AE^2 + AD_1^2 -$
$$\quad\quad 2 \times AE \times AD_1 \cos \angle D_1 AE$$

$$ED_1^2 = 64^2 + 60^2 - 2 \times 64 \times 60 \times \cos 53.13°$$

$$ED_1^2 = 7696 - 4608 = 3088$$

$$ED_1 = 55.57 \text{ mm}$$

In $\triangle BED_2$, $\angle BED_2 = 90°$, $BD_2^2 = BE^2 + D_2E^2$

$$BD_2^2 = 48^2 + 55.57^2 \quad (ED_2 = ED_1 = 55.57 \text{ mm})$$

$$BD_2^2 = 5392, BD_2 = \sqrt{5392} = 73.43 \text{ mm}$$

Also $\tan \angle BD_2E = \frac{BE}{ED_2} = \frac{48}{55.57} = 0.8638$

$$\angle BD_2E = 40.82° \ (40°49')$$

The length of BD$_2$ is 73 mm and angle BD$_2$E is 40.8°

EXERCISE 13.3

1. From a point P on the roof of a building 30 m high, the angle of elevation of A, the top of a radio mast, is 39°25'. From a window Q, 14 m above the ground, the angle of elevation of A is 50°27'. Assuming the building and the mast stand on level ground calculate the height of A, the top of the mast, above the ground.

2. The angles of elevation of the top of a tower from two points A and B are 8°27' and 15°23'. AB = 50 m and the points A, B and the foot of the tower lie in a straight line on horizontal ground. Calculate the height of the tower.

3. From a point A due north of a radio mast, the angle of elevation of the top of the mast is 11°12'. From a point B due east of the mast, the angle of elevation of the top of the mast is 17°23'. If the distance AB = 250 m, calculate the height of the mast.

4. Fig. 13.13 shows a rectangular block 50 mm by 60 mm by 70 mm. A is a point in the centre of an end face. Calculate the length of AB, and the angle between AB and the base of the block.

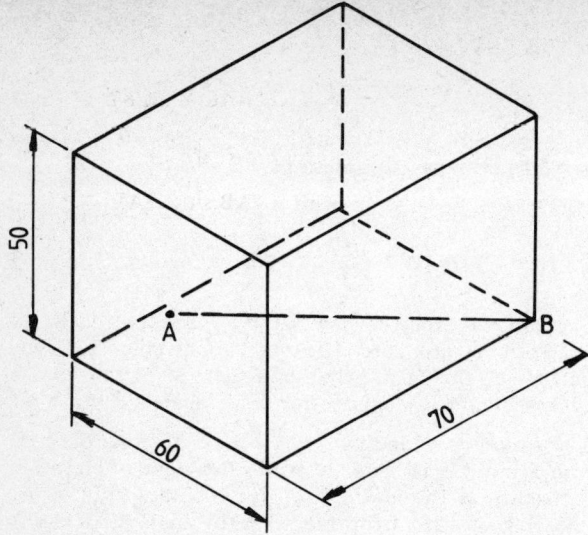

Figure 13.13

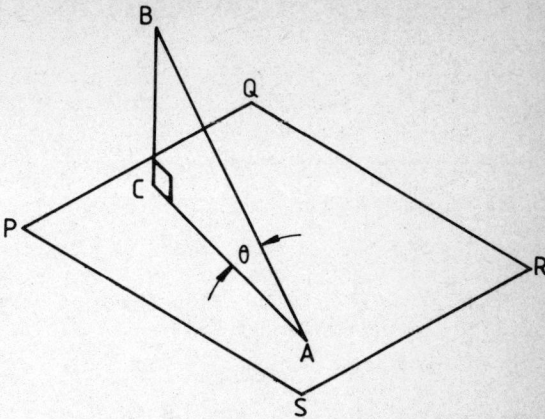

Figure 13.15

2. In fig. 13.16, the planes ABCD and ABXY intersect along AB. OP is in plane ABCD and is perpendicular to AB. OQ is in plane ABXY and is perpendicular to AB. ∠POQ = angle between the planes.

3. The angle of elevation of A from B is the angle between AB and the horizontal.

4. The angle of depression of P from Q is the angle between PQ and the horizontal.

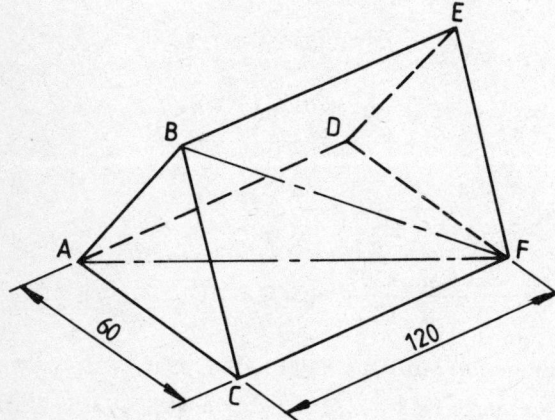

Figure 13.14

5. Fig. 13.14 shows a prism with an equilateral triangle as cross-section, each side is of length 60 mm. Calculate (a) the length of AF, (b) the length of BF, (c) the angle BFA.

6. An aircraft is flying due north and maintaining a horizontal line of flight at 600 km/h. At a certain instant a landmark is observed on a bearing of 110° at an angle of depression of 10°. One minute later the bearing of the landmark is found to be on a bearing 160°. Find the height of the aircraft.

7. A ring of radius 0.8 m is suspended by 5 strings each of length 1.2 m attached symmetrically to the ring, which hangs horizontally. The other ends of the strings being attached to a point on a beam. Find (a) the angle between adjacent strings, (b) the angle which each string makes with the horizontal.

SUMMARY

1. In fig. 13.15, the line AB meets the plane PQRS at A. BC is perpendicular to the plane. The angle θ is the angle between the line AB and the plane PQRS.

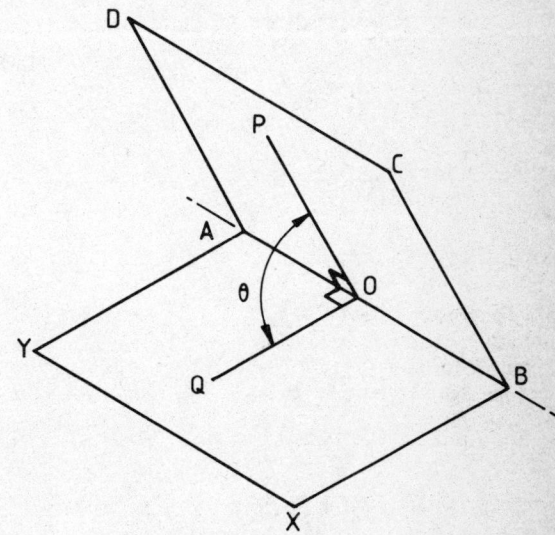

Figure 13.16

SELF ASSESSMENT PAPER No 13

Instructions: Answer all questions
Time allowed: One hour (20 marks each question)
Marks gained: 30+ pass with credit, 24–30 pass, less than 24 fail, repeat chapter 13.

1. A rectangular block of metal has a base ABCD and top PQRS. AB = CD = PQ = RS = 34.8 mm, BC = DA = PS = 23.9 mm and the height of the block is

41.8 mm. Calculate (a) the length of AC, (b) the length of AR, (c) the angle RAC.

2. A frustum of a square pyramid has a perpendicular height of 1.3 m. The square base has sides of length 1.5 m and the square top has sides 0.9 m. Calculate (a) the length of a sloping edge, (b) the angle between a sloping edge and the base, (c) the angle between a sloping face and the base.

3. A, B and C are on level ground. A mast at A is 40 m high. B is 200 m due south of A and C is east of A. If BC = 300 m calculate the angles of elevation of the top of the mast from B and C.

ANSWERS

Exercise 13.1   1. $46°13', 38°14', 18°1',$
2. $23°56',$ 3.   106 mm,   4.   76.2 mm

Exercise 13.2   1.(a) 91.9 mm, (b) $68°12',$
2. $22°2',$ 3.(a) 47.2 mm,   (b) $72°33',$
(c) $77°28',$ 4. $34°4',$ 5.   (a) $45°,$
(b) $54°44',$ (c) $109°28',$ (d) $70°31'$

Exercise 13.3   1.   64 m,   2.   16.15 m,
3.   41.8 m,   4.   80.2 mm, $18°10',$
5.(a) 134.2 mm, (b) 134.2 mm, (c) $25°50',$
6.   787 m,   7.(a) $46°9',$ (b) $48°11'.$

SELF ASSESSMENT PAPER No 13

	Marks
1. $AC^2 = 34.8^2 + 23.9^2$	5
$AC = 42.2$ mm	2
$AR^2 = 42.2^2 + 41.8^2$	5
$AR = 59.4$ mm	2
$\tan \angle RAC = \dfrac{41.8}{59.4}$	4
$\angle RAC = 35°8'$	2

2.(a) Diagonal of base $= \sqrt{1.5^2 + 1.5^2} = 2.12$ m   2
Diagonal of top $= \sqrt{0.9^2 + 0.9^2} = 1.27$ m   2
Half of difference $= \frac{1}{2}(2.12 - 1.27) = 0.425$ 2
Sloping edge $= \sqrt{0.425^2 + 1.3^2} = 1.37$ m   2

(b) Required angle $= \tan^{-1} \dfrac{1.3}{0.425}$ 4

$= 71°54'$ 2

(c) $\frac{1}{2}$(side of base − side of top) $= \frac{1}{2}(1.5 - 0.9)$
$= 0.3$ 2

Required angle $= \tan^{-1} \dfrac{1.3}{0.3}$ 2

$= 77°$ 2

3. Angle of elevation from B $= \tan^{-1} \dfrac{40}{200}$ 5

$= 11°19'$ 2

$AC^2 = 300^2 - 200^2$ 4
$AC = 223.6$ m   2

Angle of elevation from C $= \tan^{-1} \dfrac{40}{223.6}$ 5

$= 10°9'$ 2

PHASE TEST 4

Instructions: Answer all questions
Time allowed: One hour
Marks gained: 30+ pass with credit, 24−30 pass, less than 24 fail, repeat chapters 11, 12 and 13.

1. Find the vectorial sum of the currents $I_1 = 85 \sin \theta$ and $I_2 = 65 \sin \left(\theta + \dfrac{\pi}{3} \right).$

2. A tapered hole is checked by using two balls of diameters 20 mm and 40 mm. The distance between the top of the 40 mm ball and the top of the 20 mm ball is 70 mm. Calculate the taper angle of the hole.

3. A surveyor, standing at A, on a bearing of $245°$ from a radio mast at B measures the angle of elevation of the top of the mast, D, as $46°$. From a point C, on a bearing of $155°$ from the mast the angle of elevation of the top of the mast is $37°$. Determine the height of the mast if the distance between the two observation points A and C is 75 m.

ANSWERS PHASE TEST 4

	Marks
1. Let sum of vectors be $I \sin (\theta + \alpha)$	1
$I^2 = 85^2 + 65^2 - 2 \times 85 \times 65 \times \cos 120°$	4
$I^2 = 16\,975$	3
$I = 130.2$	2
$\dfrac{130.2}{\sin 120°} = \dfrac{65}{\sin \alpha}$	4
$\sin \alpha = \dfrac{65 \times \sin 120°}{130.2} = 0.4323$	3
$\alpha = 25.6°$	2
Sum of currents $I = 130.2 \sin (\theta + 25.6°)$	1

2. Let taper be θ

$\sin \dfrac{\theta}{2} = \dfrac{20 - 10}{70 - 20 + 10} = \dfrac{10}{60}$ 8

$\dfrac{\theta}{2} = 9.59°$ 4

Taper $\theta = 19.18°$ 2

3. Let BD $= h$

In $\triangle$ABD, $\tan 46° = \dfrac{h}{AB}$, $AB = \dfrac{h}{\tan 46°}$ 4

In $\triangle$BCD, $\tan 37° = \dfrac{h}{BC}$, $BC = \dfrac{h}{\tan 37°}$ 4

In $\triangle$ABC, $\angle ABC = 245° - 155° = 90°$ 4
Hence $AC^2 = AB^2 + BC^2$ 2

$75^2 = \left(\dfrac{h}{\tan 46°} \right)^2 + \left(\dfrac{h}{\tan 37°} \right)^2$ 3

$5625 = 0.9326h^2 + 1.7610h^2$ 3
$5625 = 2.6930h^2$ 2

$h = \sqrt{\dfrac{5625}{2.693}} = 45.7$ m   4

14 Arithmetic Progression, Geometric Progression and Continued Fractions

14.1 THE ARITHMETIC PROGRESSION

In an arithmetic progression each term is found by adding a constant, called the common difference, to the previous term. If the first term is a and the common difference is d the series is

$$a, a + d, a + 2d, a + 3d \ldots a + (n - 1)d$$

Example 14.1 Write down the twelfth term of 1, 4, 7, ... First term = 1, common difference $d = 4 - 1 = 7 - 4 = 3$

$$\text{Twelfth term} = a + (12 - 1)d = 1 + 11 \times 3 = 34$$

Example 14.2 A machine is bought for £450. It depreciates at the rate of £40 a year. What is the value at the end of 8 years?

At the end of the first year the value is £410   1st term
At the end of the second year the value is £370 2nd term

$$\text{Hence } a = 410, d = 370 - 410 = -40$$

At the end of eight years the value
$$= 410 + (8 - 1) \times -40 = £130$$

Example 14.3 Insert three arithmetic means between 3 and 11. This means put in three numbers so that the five numbers form an arithmetic progression.

$$\text{First term} = 3, \text{Fifth term} = a + 4d = 11$$

$$\text{Hence } 3 + 4d = 11, 4d = 8, d = 2$$

The required numbers are $3 + 2, 3 + 2 \times 2, 3 + 3 \times 2$

$$\text{or } 5, \quad 7, \quad 9$$

If we write l for the last term of an AP, that is $l = a + (n - 1)d$ then the sum of n terms, S_n can be written

$$S_n = a + (a + d) + (a + 2d) + \ldots + (l - 2d) + (l - d) + l \tag{1}$$

or

$$S_n = l + (l - d) + (l - 2d) + \ldots + (a + 2d) + (a + d) + a \tag{2}$$

If we add we have
$$2S_n = (a + l) + (a + l) \ldots + (a + l) + (a + l) + (a + l)$$
or $2S_n = n(a + l)$ since there are n terms

$$\text{Hence } S_n = \frac{n}{2}(a + l)$$

But $l = a + (n - 1)d$, so $S_n = \frac{n}{2}(a + a + [n - 1]d)$

Hence the sum of an AP to n terms is given by S_n where

$$S_n = \frac{n(a + l)}{2}, \text{ or } S_n = \frac{n}{2}(2a + [n - 1]d)$$

Example 14.4 Find the sum of the first twenty-eight terms of the series 5, 7.5, 10, ...

$$a = 5, d = 2.5, n = 28$$

$$S_{28} = \frac{28}{2}(2 \times 5 + [28 - 1] \times 2.5)$$

$$S_{28} = 14(10 + 67.5) = 1085$$

Example 14.5 How many terms of 76, 73, 70 ... will give a sum of 714?

$$a = 76, d = -3, n = ?, S_n = 714$$

$$714 = \frac{n}{2}(2 \times 76 + [n - 1] \times -3)$$

$$714 = 76n - \frac{3n^2}{2} + \frac{3n}{2}$$

Multiply by 2 and rearrange

$$3n^2 - 155n + 1428 = 0$$

$$(3n - 119)(n - 12) = 0$$

$$3n = 119 \text{ or } n = 12$$

Twelve terms total 714, the $39\frac{2}{3}$ terms means that the AP has been taken until the terms become negative and hence the sum begins to decrease.

EXERCISE 14.1

1. Find the eighth term of 65, 62, 59, _ _ _
2. Find the fifth term of 6, 11, 16, _ _ _
3. Find the twenty-first term of 24½, 23, 21½, _ _ _

4. If the seventh term of an AP is 15 and the twelfth term is 17½ find the first term.

5. Insert three arithmetic means between 32 and 20.

6. If the fifth term of an AP is 1.3 and the fifteenth term is 2.8 find the third term.

7. Find the sum of the first twenty four terms of 1, 7, 13, _ _ _

8. Find the sum of the first eighteen terms of 17, 16.8, 16.6, _ _ _

9. Sum, 5, 6.2, 7.4, _ _ _ to twenty two terms.

10. How many terms of the series 3, 3.1, 3.2, _ _ _ total 89.1?

11. Which term of the series 1.5, 1.8, 2.1, _ _ is 6.9?

14.2 THE GEOMETRIC PROGRESSION

In a geometric progression each term is derived by multiplying the previous term by a constant called the common ratio. If the first term is a, and the common ratio is r the series is $a, ar, ar^2, ar^3, \ldots ar^{n-1}$.

Example 14.6 What is the seventh term of $1, 3, 9, \ldots$?

$$a = 1, ar = 3, \text{ i.e. } r = 3$$
$$\text{Seventh term} = ar^{7-1} = ar^6 = 1 \times 3^6 = 729$$

Example 14.7 The third term of a GP is $\dfrac{2}{25}$ and the fourth term is $-\dfrac{2}{125}$. Find the first term.

$$\text{third term } ar^2 = \frac{2}{25}, \text{ fourth term } ar^3 = -\frac{2}{125}$$

$$r = \frac{ar^3}{ar^2} = -\frac{2}{125} \times \frac{25}{2} = -\frac{1}{5}, r^2 = \frac{1}{25}$$

But $ar^2 = \dfrac{2}{25}$, hence $a \times \dfrac{1}{25} = \dfrac{2}{25}, a = 2$

The first term is 2

Example 14.8 Insert two geometric means between 7.3 and 21.5. This means that two numbers are required between 7.3 and 21.5 such that the four numbers form a geometric progression.
first term $a = 7.3$, fourth term $ar^3 = 21.5$

$$r^3 = \frac{21.5}{7.3} = 2.95$$
$$r = \sqrt[3]{2.95} = 1.43$$

The geometric means are 7.3×1.43 and 7.3×1.43^2
The geometric means are 10.5 and 15.0
If the sum of n terms of a GP is S_n then

$$S_n = a + ar + ar^2 + \ldots + ar^{n-1} \qquad (1)$$
$$rS_n = ar + ar^2 + ar^3 + \ldots + ar^{n-1} + ar^n \qquad (2)$$

Subtract (1) from (2) $rS_n - S_n = ar^n - a$

all of the other terms cancel

$$S_n(r-1) = a(r^n - 1)$$
$$S_n = \frac{a(r^n - 1)}{r - 1}$$

When r is greater than 1 the above formula is used. When r is less than 1 the formula is written

$$S_n = \frac{a(1 - r^n)}{1 - r}$$

Example 14.9 Find the sum of the first eight terms of $3, 1, \frac{1}{3}, \ldots$

$$a = 3, r = \frac{1}{3}, n = 8$$

$$S_8 = \frac{3(1 - [\frac{1}{3}]^8)}{1 - \frac{1}{3}} = \frac{3(1 - 0.000\,15)}{\frac{2}{3}}$$

$$S_8 = 4.50$$

Example 14.10 The rate of drilling an oil well decreases by 10% every hour. If the number of metres drilled in the first hour is 16, find the depth at the end of 6 hours and also how long it will take to drill to a depth of 90 m. The number of metres drilled per hour forms a GP with $a = 16$ and $r = \dfrac{9}{10}$

$$\text{Depth drilled at the end of 6 hours} = \frac{16\left(1 - \left[\dfrac{9}{10}\right]^6\right)}{1 - \dfrac{9}{10}}$$

$$= \frac{16(1 - 0.53)}{\dfrac{1}{10}}$$

$$= 75.2 \text{ m}$$

Let the number of hours required to drill 90 m be n.

$$\text{Then } 90 = \frac{16\left(1 - \left[\dfrac{9}{10}\right]^n\right)}{1 - \dfrac{9}{10}}$$

$$90 = 160 - 160\left(\frac{9}{10}\right)^n$$

$$160(0.9)^n = 70$$

$$0.9^n = \frac{70}{160} = 0.4375$$

Take $\log_{10}$ of both sides $\log_{10} 0.9^n = \log_{10} 0.4375$

$$n \log_{10} 0.9 = \log_{10} 0.4375$$

$$n = \frac{\log_{10} 0.4375}{\log_{10} 0.9} = \frac{-0.3590}{-0.0458} = 7.85$$

It takes 7.85 hours to drill 90 m

In *Example 14.9* the term $(\frac{1}{3})^8$ was negligible compared with 1. In general if r is less than 1 and n is large then r^n can be ignored. The sum is then called the sum to infinity S_∞ and it means the sum when n is very large.

$$S_\infty = \frac{a}{1-r}$$

Example 14.11 Find the sum to infinity of $6, -1, \frac{1}{6}, \ldots$

$$a = 6, r = -\frac{1}{6}$$

$$S_\infty = \frac{6}{1 - \left(-\frac{1}{6}\right)} = \frac{6}{\frac{7}{6}} = 5\frac{1}{7}$$

Example 14.12 In *Example 14.10* find the maximum depth that can be drilled.

Maximum depth will be the sum to infinity,

$$S_\infty = \frac{16}{1 - \frac{9}{10}} = 160\,\text{m}$$

EXERCISE 14.2

1. Find the seventh term of $1, 5, 25, \_\_\_$

2. Find the eleventh term of $2, 2.2, 2.42, \_\_\_$

3. If the seventh term of a GP is 384 and the ninth term is 1536, find the fourth term.

4. If the fifth term of a GP is $\frac{1}{3}$ and the eighth term is $\frac{1}{81}$, find the eleventh term.

5. Insert two geometric means between 3.2 and 10.8.

6. Find the sum of the first eight terms of $2.4, 0.6, 0.15, \_\_\_$

7. How many terms of the series $\frac{1}{4}, \frac{1}{2}, 1 \_\_\_$ will total 63¾?

8. Find the sum to infinity of $1, \frac{1}{5}, \frac{1}{25}, \_\_\_$

9. A ball is dropped from a height of 8 m on to a horizontal floor and bounces repeatedly, the heights of successive rebounds being in GP. If the height of the eighth bounce is ½ m, find the total distance the ball will travel before it comes to rest.

10. The gear speeds in rev/min of a drilling machine form a GP. If in first gear the speed is 120 rev/min, and in the fourth gear 405 rev/min find the speeds of the second and third gears.

14.3 CONTINUED FRACTIONS

To find the convergents for a ratio which has a higher number in the denominator the following procedure is used.

Example 14.13 Find the convergents for the ratio $\frac{47}{127}$

```
47)127(2
   94
   33)47(1
      33
      14)33(2
         28
         5)14(2
           10
           4)5(1
             4
             1)4(4
```

As can be seen above the procedure is:

Divide the numerator (47) into the denominator (127)

Divide the remainder (33) into the first divisor (47)

Divide the remainder (14) into the second divisor (33) etc.

From the above the numbers 2, 1, 2, 2, 1, 4 are called the quotients. A table is now formed to calculate the convergents.

	1	2	2	1	4
0	1	1	3	7	10	47
1	2	3	8	19	27	127
	1	2	2	1	4

The convergents are $\frac{1}{2}, \frac{1}{3}, \frac{3}{8}, \frac{7}{19}, \frac{10}{27}, \frac{47}{127}$ and these have been obtained as follows.

First write down $\frac{0}{1}$ on the left of the table, then the first convergent is 1 divided by the first quotient 2, that is $\frac{1}{2}$

The other convergents are then obtained as follows

$$\frac{0}{1}\ \frac{1}{2}\ \frac{\overset{1}{1 \times 1 + 0}}{\underset{1}{1 \times 2 + 1}} = \frac{1}{3}$$

$$\frac{1}{2}\ \frac{1}{3}\ \frac{\overset{2}{2 \times 1 + 1}}{\underset{2}{2 \times 3 + 2}} = \frac{3}{8}$$

$$\frac{1}{3}\ \frac{3}{8}\ \frac{\overset{2}{2 \times 3 + 1}}{\underset{2}{2 \times 8 + 3}} = \frac{7}{19}\ \text{etc.}$$

The convergents are approximations to the given fraction, each one being closer to the given fraction.

Fraction	$\dfrac{1}{2}$	$\dfrac{1}{3}$	$\dfrac{3}{8}$	$\dfrac{7}{19}$
Decimal	0.5	0.333 33	0.375	0.368 42
Error	0.129 92	− 0.036 75	0.004 92	− 0.001 66

Fraction	$\dfrac{10}{27}$	$\dfrac{47}{127}$
Decimal	0.370 37	0.370 08
Error	0.000 29	0

The convergents are $\dfrac{1}{2}, \dfrac{1}{3}, \dfrac{3}{8}, \dfrac{7}{19}, \dfrac{10}{27}, \dfrac{47}{127}$

When the fraction is such that the numerator is larger than the denominator the fraction is inverted and the convergents found for the new fraction. These convergents are then inverted to obtain the required convergents.

Example 14.14 Find the convergents for the ratio $\dfrac{131}{107}$.

Write the ratio as $\dfrac{107}{131}$.

```
107)131(1
    107
     24)107(4
         96
         11)24(2
             22
              2)11(5
                10
                11)2(2
```

		4	2	5	2
0	1	4	9	49	107
1	1	5	11	60	131
		4	2	5	2

Required convergents are $\dfrac{1}{1}, \dfrac{5}{4}, \dfrac{11}{9}, \dfrac{60}{49}, \dfrac{131}{107}$

EXERCISE 14.3

Find the convergents for the following fractions
1. $\dfrac{19}{61}$, 2. $\dfrac{43}{91}$, 3. $\dfrac{393}{450}$, 4. $\dfrac{1250}{319}$, 5. $\dfrac{250}{433}$
6. Write 0.87 as a fraction and hence determine the first four convergents.

14.4 APPLICATIONS OF CONTINUED FRACTIONS

Example 14.15 Obtain the closest approximation to $\dfrac{53}{89}$ that can be obtained from a set of change gear wheels ranging from 20T to 120T in steps of 5T.

```
53)89(1
   53
   36)53(1
      36
      17)36(2
         34
          2)17(8
             16
              1)2(2
```

	1	2	8	2
0	1	1	3	25	53
1	1	2	5	42	89
	1	2	8	2

Closest approximation is $\dfrac{25}{42} = \dfrac{5 \times 5}{7 \times 6} = \dfrac{50}{70} \times \dfrac{25}{30}$

Hence the closest approximation is obtained by using

$$\dfrac{50T}{70T} \times \dfrac{25T}{30T}$$

$\dfrac{53}{89} = 0.595\,506, \dfrac{25}{42} = 0.595\,238$

The error is $0.595\,506 - 0.595\,238 = 0.000\,268$ a 0.045% error

Example 14.15 Determine the convergents for the ratio $\dfrac{457}{560}$.

Show that $\dfrac{p_4 + p_5}{q_4 + q_5}$ gives a fraction which factorises to $\dfrac{6 \times 17}{5 \times 25}$.

```
457)560(1
    457
    103)457(4
        412
        45)103(2
           90
           13)45(3
              39
               6)13(2
                 12
                  1)6(6
```

		4	2	3	2	6
0	1	4	9	31	71	457
1	1	5	11	38	87	560
		4	2	3	2	6

$$\frac{p_4 + p_5}{q_4 + q_5} = \frac{31 + 71}{38 + 87} = \frac{102}{125} = \frac{6 \times 17}{5 \times 25}$$

EXERCISE 14.4

1. An external screw thread, having a lead of 2.65 mm, is to be cut on a precision instrument lathe equipped with a lead-screw of 3 mm lead, and a set of change gear wheels ranging from 20T to 100T in steps of 5T. Determine the nearest lead that can be cut on the lathe. Given

$$\frac{\text{Drivers}}{\text{Driven}} = \frac{\text{Lead of thread to be cut}}{\text{Lead of leadscrew}}$$
$$= \frac{2.65}{3} = \frac{53}{60}$$

2. A screw thread having a lead of 7.40 ± 0.02 mm is to be cut on a lathe having a leadscrew of 10 mm lead and a set of change gear wheels from 20T to 120T in steps of 5T. Determine suitable change gear wheels.

3. A thread of pitch 2.5 mm is to be cut on a lathe having a leadscrew of 6.35 mm lead. Obtain a continued fraction for the ratio formed. No single wheel of the train has to have more than 100 teeth. Show that a suitable compound gear train is provided by the ratio $\dfrac{p_5 + p_6}{q_5 + q_6}$ where $\dfrac{p_n}{q_n}$ represents the nth convergent. If this ratio is used find the cumulative pitch error over a length of 100 pitches.

4. In hobbing a gear the ratio $\dfrac{3926}{2741}$ is required. Express the ratio as a continued fraction and show that

$$\frac{2p_6 + 5p_7}{2q_6 + 5q_7} = \frac{59}{31} \times \frac{73}{97}$$

where $\dfrac{p_n}{q_n}$ is the nth convergent.

14.5 GENERAL EXAMPLES

Example 14.17 A dividing head has a 40 to 1 worm gear ratio so that one turn of the operating handle rotates the main spindle by 9 degrees. A particular setting requires the main spindle to rotate through an angle of $6°33'$. The dividing head index plate has every hole circle from 30 holes to 60 holes inclusive. Determine the nearest approximate hole spacing and hole circle that is revealed by the method of continued fractions. Find also the error in seconds.

$$6°33' = 393' \text{ and } 9° = 540'$$

$$\text{Turn of the index arm} = \frac{6°33'}{9°}$$
$$= \frac{393}{540}$$

```
393)540(1
    393
    147)393(2
        294
         99)147(1
             99
             48)99(2
                96
                 3)48(16
```

		2	1	2	16
0	1	2	3	8	131
1	1	3	4	11	180
		2	1	2	16

The nearest fraction is $\dfrac{8}{11}$ which is equivalent to $\dfrac{24}{33}$ or $\dfrac{32}{44}$ or $\dfrac{40}{55}$. This is equivalent to 24 holes in a 33 hole circle.
or 32 holes in a 44 hole circle.
or 40 holes in a 55 hole circle.

$$\text{Angular spacing} = \frac{8}{11} \times 9° = 6°32'44''$$

$$\text{The error} = 6°33' - 6°32'44'' = 16 \text{ seconds}$$

EXERCISE 14.5

1. (a) If the first term of an AP is 27 and the nineteenth term is 43 find the common difference and the sum of nineteen terms.

(b) Find the sum to infinity of the series: 60, 48, 38.4 _ _ _

(c) A machine costs £10 000 when new. It is estimated that it depreciates in value by 10% each year. In how many years, to the nearest year, will it be down to its scrap value of £100?

2. (a) Find five numbers in AP such that their sum is 155 and the difference between the first and last is 28.

(b) The amplitudes of successive swings of a pendulum are in GP. The amplitudes of the first and second swings are 250 mm and 200 mm respectively. What is the amplitude of the tenth swing? Find also the sum of the amplitudes of the first 18 swings.

3. A milling machine has a machine table leadscrew of 6 mm lead and is equipped with a dividing head which has 40 to 1 worm and wormwheel ratio. Change gear wheels available are 17, 18, 19, 20, 21, 22, 24, 27, 30, 33, 36, 42, 48, 51, 55 and 60 teeth. Find the change gear wheels that can be used to cut a helix of $\dfrac{211.02}{210.98}$ mm leads.

4. Determine the convergents for the ratio $\dfrac{57\ 351}{81\ 982}$. If the nth convergent is $\dfrac{p_n}{q_n}$, show that $\dfrac{2p_6 + p_4}{2q_6 + q_4}$ gives a compound gear train suitable for a hobbing machine.

SUMMARY

1. The terms of an AP are $a, a + d, a + 2d \ldots a + (n - 1)d$, a is the first term and d is the common difference.

2. The sum of n terms of an AP is given by

$$S_n = \frac{n}{2}(2a + [n-1]d) \text{ or } S_n = \frac{n(a+l)}{2}$$

where l is the last term.

3. The terms of a GP are $a, ar, ar^2, \ldots ar^{n-1}$, a is the first term and r is the common ratio.

4. The sum of n terms of a GP is given by

$$S_n = \frac{a(1-r^n)}{1-r} \text{ when } r < 1 \text{ and}$$

$$S_n = \frac{a(r^n-1)}{r-1} \text{ when } r > 1.$$

5. When $r < 1$ and n is large the sum becomes the sum to infinity and $S_\infty = \frac{a}{1-r}$.

6. Continued fractions is a method of determining a set of convergent fractions and is used to find an approximate simple fraction to a complicated fraction.

7. In some cases the convergents can be combined to find a better approximation.

8. Continued fractions are used in selection of gear wheels for screw cutting, for determining milling machine change gear ratios, determining angular spacing using a dividing head and in hobbing.

SELF ASSESSMENT PAPER No 14

Instructions: Answer all questions in both sections
Time allowed: Section A 30 minutes (50 marks)
Section B 1 hour (30 marks each question)
Marks gained: 55+ pass with credit, 44–55 pass, less than 44, fail, repeat chapter 14

Section A

1. Find the nineteenth term of $7, 6.6, 6.2, ---$

2. Insert two arithmetic means between 17 and 24.8

3. Find the ninth term of $8, 2, \frac{1}{2}, ---$

4. Find the sum of the first ten terms of $7, 14, 28, ---$

5. Find the sum to infinity of $0.2, 0.06, 0.018, ---$

6. Find the convergents for the ratio $\frac{29}{83}$.

7. Find the convergents for the ratio $\frac{97}{31}$.

Section B

1. (a) The first three terms of a progression are $1, 1\frac{3}{8}, 1\frac{3}{4}$. Determine (i) the ninth term, (ii) the sum of the first nine terms.

(b) A lathe is designed to accommodate cylindrical objects having diameters from 50 mm to 150 mm. The cutting speed i.e. the speed of a point on the surface of the object being turned, is to be 22.5 metres per minute. Six speeds in GP are required to cover this range of work. Find these speeds, correct to the nearest rev/min.

2. In cutting a lead on a milling machine the following gear ratio is required $\frac{719}{500}$. The change wheels available are: 17, 18, 19, 20, 21, 22, 24, 27, 30, 33, 36, 39, 42. Show that $\frac{p_6 - p_2}{q_6 - q_2}$ gives a suitable ratio for a compound gear train where $\frac{p_n}{q_n}$ is the nth convergent.

ANSWERS

Exercise 14.1 1. 44, 2. 26, 3. $-5\frac{1}{2}$,
4. 12, 5. 29, 26, 23, 6. 1,
7. 1680, 8. 275.4, 9. 387.2,
10. 22, 11. 19th.

Exercise 14.2 1. 15 625, 2. 5.2, 3. ± 48,
4. $\frac{1}{2187}$, 5. 4.80, 7.23, 6. 3.2,
7. 8, 8. $1\frac{1}{4}$, 9. 27.3 m, 10. 180, 270.

Exercise 14.3 1. $\frac{1}{3}, \frac{4}{13}, \frac{5}{16}$, 2. $\frac{1}{2}, \frac{8}{17}, \frac{9}{19}, \frac{17}{36}$,

3. $\frac{1}{1}, \frac{6}{7}, \frac{7}{8}, \frac{62}{71}$, 4. $\frac{3}{1}, \frac{4}{1}, \frac{47}{12}, \frac{145}{37}, \frac{192}{49}, \frac{529}{135}$,

5. $\frac{1}{1}, \frac{1}{2}, \frac{3}{5}, \frac{4}{7}, \frac{11}{19}, \frac{15}{26}, \frac{41}{71}, \frac{56}{97}, \frac{97}{168}$, 6. $\frac{1}{1}, \frac{6}{7}$,

$\frac{7}{8}, \frac{20}{23}$.

Exercise 14.4 1. $\frac{75T}{85T}$, lead 2.647 mm,

2. $\frac{85T}{115T}$, 3. $\frac{63}{160} = \frac{35T \times 45T}{50T \times 80T}$ error 0.031 25 mm

Exercise 14.5 1 (a) $\frac{8}{9}$, 665, (b) 300, (c) 44,

2. (a) 45, 38, 31, 24, 17 (b) 33.5 mm, 1227.5 mm,

3. $\frac{24T \times 30T}{21T \times 39T}$ or $\frac{48T \times 30T}{42T \times 39T}$ 4. $\frac{1}{1}, \frac{2}{3}, \frac{7}{10}$,

$\frac{156}{223}, \frac{631}{902}, \frac{1418}{2027}$

SELF ASSESSMENT PAPER No 14

Marks

Section A
1. $a = 7, d = -0.4, n = 19$ 3
$a + (n-1)d = -0.2$ 2

	Marks
2. $17 + 3d = 24.8$, $d = 2.6$,	3
19.6, 22.2	4
3. $a = 8$, $r = \frac{1}{4}$, $n = 9$	3
$ar^{n-1} = 8 \times (\frac{1}{4})^8 = 0.000\,12 \left(\dfrac{1}{8192}\right)$	4
4. $a = 7$, $r = 2$, $n = 10$	3
$S_{10} = \dfrac{7(2^{10} - 1)}{2 - 1} = 7161$	4
5. $a = 0.2$, $r = 0.3$	2
$S_\infty = \dfrac{0.2}{1 - 0.3} = 0.286 \,(\frac{2}{7})$	3
6. Quotients 2, 1, 6, 4	4
Convergents $\dfrac{1}{2}, \dfrac{1}{3}, \dfrac{7}{20}, \dfrac{29}{83}$	4
7. Write ratio $\dfrac{31}{97}$	1
Quotients 3, 7, 1, 3	4
Convergents $\dfrac{1}{3}, \dfrac{7}{22}, \dfrac{8}{25}, \dfrac{31}{97}$	4
Inverted convergents $\dfrac{3}{1}, \dfrac{22}{7}, \dfrac{25}{8}, \dfrac{97}{31}$	2

Section B

	Marks
1. (a) $a = 1$, $d = \frac{3}{8}$	2
(i) $n = 9$, $a + (n - 1)d = 4$	2
(ii) $S_9 = \frac{9}{2}(2 + 8 \times \frac{3}{8})$ or $\dfrac{9(1 + 4)}{2}$	4
$s_9 = 22.5$	2
(b) $a = 50$, $ar^5 = 150$	2
$r^5 = \dfrac{150}{50} = 3$	3
$r = 1.246$	2
Diameters are 50, 62.3, 77.6, 96.7, 120.5, 150 mm	6
Speed (rev/min) $= \dfrac{22\,500}{\text{circumference}}$	2
Speeds are $\dfrac{22\,500}{50}, \dfrac{22\,500}{62.3}$ etc.	2
Speeds are 144, 116, 92, 74, 60, 48 rev/min	3
2. Invert ratio to $\dfrac{500}{719}$	2
Quotients 1, 2, 3, 1, 1, 7, 4	8
Convergents $\dfrac{1}{1}, \dfrac{2}{3}, \dfrac{7}{10}, \dfrac{9}{13}, \dfrac{16}{23}, \dfrac{121}{174}, \dfrac{500}{719}$	8
Inverting the convergents	
$\dfrac{1}{1}, \dfrac{3}{2}, \dfrac{10}{7}, \dfrac{13}{9}, \dfrac{23}{16}, \dfrac{174}{121}, \dfrac{719}{500}$	2
$\dfrac{p_6 - p_2}{q_6 - q_2} = \dfrac{174 - 3}{121 - 2} = \dfrac{171}{119}$	4

	Marks
$\dfrac{171}{119} = \dfrac{19T \times 27T}{17T \times 21T}$	6

EXAMINATION 2

Instructions: Answer all questions in Section A and 4 questions from Section B

Time allowed: Section A 45 minutes (40 marks)
Section B 75 minutes
(15 marks each question)

Marks gained: 50+ pass with credit, 40–50 pass, less than 40 fail, repeat chapters 8–14.

Section A

1. Find the tenth term and the sum of the first ten terms of the series 7.5, 5.0, 2.5, _ _ _

2. In $\triangle ABC$, $AB = 40$ mm, $BC = 20$ mm and $\angle ABC = 90°$. The triangle is rotated through $360°$ about the side AB. Use the theorem of Pappus to calculate the volume of the cone formed.

3. Find the volume of a regular pyramid whose base is a square of side 20 mm and whose perpendicular height is 60 mm.

4. An alternating current has the following values.

Time (ms)	0	5	10	15	20	25	30
Current (A)	0	0.68	2.00	3.68	4.35	2.63	0

Use Simpson's Rule to find the area under the curve formed and hence determine the mean value.

5. Given in $\triangle ABC$ $a = 2$ m, $b = 3$ m and $C = 47°$. Calculate (i) the value of the side c and (ii) the area of the triangle.

Section B

1. (a) The second term of an AP is 8.1. The fourth term is 10.9. Determine the sum of the first 25 terms.

(b) The pitches of BA threads are in GP. Size 2BA is 0.81 mm and size 3BA is 0.729 mm. Find the pitches of sizes 1BA and 4BA.

(c) A ball when dropped from a height of h metres on to a hard level surface rebounds to a height of $0.64h$ metres. Calculate the total distance travelled by the ball before coming to rest if it is dropped from a height of 2.7 m.

2. A table leadscrew of a milling machine has a lead of 6 mm, the dividing head available has a worm and worm-wheel ratio of 40 to 1 and the following set of change gear wheeels are also available, 17, 18, 19, 20, 21, 22, 24, 27, 30, 33, 36, 39, 42, 45, 48, 51, 55, and 60 teeth. Show that a suitable change gear ratio for cutting a helix of lead 163 ± 0.2 mm can be obtained by considering the ratio $\dfrac{p_4 + p_5}{q_4 + q_5}$ where $\dfrac{p_n}{q_n}$ is the nth convergent of the ratio formed.

3. A solid metal frustum of a cone has end diameters 80 mm and 240 mm and height 120 mm. Calculate the volume of the frustum and its mass if the metal has a density of 2.7×10^3 kg/m³. If the frustum is recast as a sphere, without loss of material, calculate its radius.

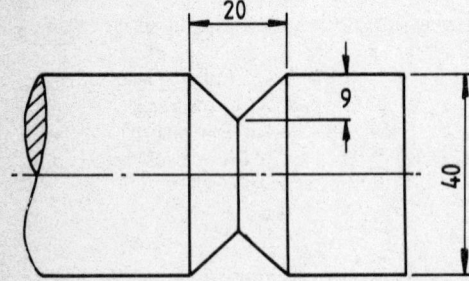

Figure 1

4. A 40 mm diameter steel bar has a symmetrical triangular groove machined to the dimensions given in fig. 1, all dimensions are in mm. Calculate the volume of the metal removed in cm³.

5. Three holes P, Q, R are drilled in a plate. PQ = 50.5 mm, PR = 100.8 mm and ∠QPR = 110°27′. Calculate the length of RQ and the angle PRQ.

6. A cube of side 2 m has base ABCD and vertical edges AE, BF, CG, DH. Calculate (a) the length of diagonal BH, (b) the angle BH makes with the base ABCD.

ANSWERS EXAMINATION 2

Marks

Section A

1. $a = 7.5, d = -2.5$ 1
10th term $= a + 9d = 7.5 + 9 \times -2.5$ 2
$\qquad = -15$ 1

$S_{10} = \dfrac{10}{2}(7.5 - 15)$ or

$S_{10} = \dfrac{10}{2}(2 \times 7.5 + 9 \times -2.5)$ 3

$\qquad = -37.5$ 1

2. Area of $\triangle ABC = \frac{1}{2} \times 40 \times 20 = 400$ 2

Distance of centroid from AB $= \frac{1}{3} \times BC = \dfrac{20}{3}$ 2

Volume $= 400 \times 2\pi \times \dfrac{20}{3}$ 3

$\qquad = 16\,760$ mm³   1

3. Volume $= \frac{1}{3} \times 20 \times 20 \times 60$ 4
$\qquad = 8000$ 1

4. Area $= \frac{5}{3}(0 + 0 + 4[0.68 + 3.68 + 2.63]$
$\qquad + 2[2.00 + 4.35])$ 4
$\qquad = 67.77$ 2

Mean value $= \dfrac{67.77}{30} = 2.26$ A   2

5. (a) $c^2 = 2^2 + 3^2 - 2 \times 2 \times 3 \times \cos 47°$ 4
$\qquad c = \sqrt{4.816} = 2.195$ m   2
  (b) Area $= \frac{1}{2} \times 2 \times 3 \times \sin 47°$ 3
$\qquad = 2.194$ m²   2

Section B

1. (a) $a + d = 8.1, a + 3d = 10.9$ 1
$2d = 10.9 - 8.1, d = 1.4, a = 6.7$ 2

$S_{25} = \dfrac{25}{2}(2 \times 6.7 + 24 \times 1.4)$ 2

$\qquad = 587.5$ 1
(b) $ar = 0.81, ar^2 = 0.729$ 1

$r = \dfrac{0.729}{0.81} = 0.9, a = 0.9$ 2

1BA = 0.90 mm, 4BA = 0.656 mm   2
(c) $a = 2.7$ m, $r = 0.64$ 2

$S_\infty = \dfrac{2.7}{1 - 0.64} = 7.5$ m   2

2. Lead of machine $= 40 \times 6 = 240$ mm   1
$\dfrac{\text{Driven}}{\text{Drivers}} = \dfrac{\text{lead of helix to be cut}}{\text{lead of machine}} = \dfrac{163}{240}$ 2
Quotients are 1, 2, 8, 1, 1, 4   3
Convergents are $\dfrac{1}{1}, \dfrac{2}{3}, \dfrac{17}{25}, \dfrac{19}{28}, \dfrac{36}{53}, \dfrac{163}{240}$ 3

$\dfrac{p_4 + p_5}{q_4 + q_5} = \dfrac{19 + 36}{28 + 53} = \dfrac{55}{81}$ 2

$\dfrac{\text{Driven}}{\text{Drivers}} = \dfrac{5 \times 11}{9 \times 9} = \dfrac{20T \times 33T}{36T \times 27T}$ 2

Lead of helix produced $= \dfrac{55}{81} \times 240 = 162.96$ mm   1
This is within limits of 163 ± 0.2 mm   1

3. Volume of frustum $= \dfrac{\pi \times 120}{3} \times$
$\qquad (40^2 + 120^2 + 40 \times 120)$ 4
$\qquad = 2.614 \times 10^6$ mm³   2

Mass $= \dfrac{2.614 \times 10^6}{10^9} \times 2.7 \times 10^3$ 2

$\qquad = 7.06$ kg   3
$\frac{4}{3}\pi R^3 = 2.614 \times 10^6$ 2
$R^3 = \dfrac{3 \times 2.614 \times 10^6}{4} = 624 \times 10^3$ 2
$R = 85.45$ mm   1

4. Area of $\triangle = \frac{1}{2} \times 20 \times 9 = 90$ mm²   4
Distance of centroid from centre of bar
$\qquad = 20 - \frac{1}{3} \times 9$
$\qquad = 17$ mm   4
Volume removed $= 90 \times 2\pi \times 17$ 4
$\qquad = 9613$ mm³ $= 9.613$ cm³   3

	Marks		Marks

5. Length of RQ is found by the cosine rule   **1**

$$RQ^2 = 100.8^2 + 50.5^2 - 2 \times 100.8 \times 50.5$$
$$\times \cos 110°27'$$ **4**

$$RQ^2 = 12\,711 + 3557$$ **2**

$$RQ = \sqrt{16\,268} = 127.5 \text{ mm}$$ **2**

$$\frac{127.5}{\sin 110°27'} = \frac{50.5}{\sin \angle PRQ}$$ **2**

$$\sin \angle PRQ = \frac{50.5 \times \sin 110°27'}{127.5} = 0.3711$$ **2**

$\angle PRQ = 21.78°\ (21°47')$ **2**

6. Diagonal of base $BD = \sqrt{2^2 + 2^2} = 2.828$ **4**

In $\triangle BDH$, $BH^2 = 2.828^2 + 2^2$ **4**

$BH = \sqrt{11.998} = 3.46 \text{ m}$ **2**

In $\triangle BDH$, $\sin \angle HBD = \dfrac{2}{3.46} = 0.5774$ **3**

$\angle HBD = 35.27°\ (35°16')$ **2**

15 Quadratic Equations

15.1 QUADRATIC EXPRESSIONS

If two linear expressions, that is two expressions which contain only an x term and a constant term, are multiplied together they form a quadratic expression. Thus

$$(x + 3)(2x + 1) = x(2x + 1) + 3(2x + 1)$$
$$= 2x^2 + x + 6x + 3$$
$$= 2x^2 + 7x + 3$$

If we reverse the process we say that the factors of

$$2x^2 + 7x + 3 \text{ are } (x + 3)(2x + 1)$$

A quadratic expression has an x^2 term in it, it may also have an x term or a constant term but it will not have any powers of x above 2. e.g. $3x^2 + 6x - 1$, $2x^2 + 6$, $3x^2 + 7x$ are all quadratic expressions.
Some quadratic expressions have factors but some do not have factors.

Example 15.1 Factorise the following, where factors exist.

(a) $3x^2 + 7x$, (b) $x^2 - 9$, (c) $x^2 + 5x + 6$,
(d) $x^2 - x - 6$, (e) $x^2 + 2x + 5$, (f) $8x^2 - 14x - 15$,
(g) $4x^2 - 20x + 25$, (h) $x^2 + 9$

(a) $3x^2 + 7x = x(3x + 7)$. The factors are x and $3x + 7$

(b) $x^2 - 9 = (x - 3)(x + 3)$. The factors are $x - 3$ and $x + 3$

In general $a^2 - b^2 = (a - b)(a + b)$

(c) $x^2 + 5x + 6 = x^2 + 3x + 2x + 6$
$$= x(x + 3) + 2(x + 3)$$
$$= (x + 2)(x + 3)$$

(d) $x^2 - x - 6 = x^2 - 3x + 2x - 6$
$$= x(x - 3) + 2(x - 3)$$
$$= (x + 2)(x - 3)$$

(e) $x^2 + 2x + 5$ has no factors

(f) $8x^2 - 14x - 15 = 8x^2 + 6x - 20x - 15$
$$= 2x(4x + 3) - 5(4x + 3)$$
$$= (2x - 5)(4x + 3)$$

(g) $4x^2 - 20x + 25 = 4x^2 - 10x - 10x + 25$
$$= 2x(2x - 5) - 5(2x - 5)$$
$$= (2x - 5)(2x - 5)$$

$4x^2 - 20x + 25$ is known as a perfect square, that is the factors are equal.
In general $(a + b)^2 = a^2 + 2ab + b^2$ and $(a - b)^2 = a^2 - 2ab + b^2$

(h) $x^2 + 9$ has no factors.

EXERCISE 15.1

1. Multiply out the following brackets: (a) $2x(x - 3)$, (b) $(2x - 1)(x + 5)$, (c) $(x - 7)(5x + 2)$, (d) $(3x + 4)^2$, (e) $(3x - 5)(3x + 5)$, (f) $(7x - 3)(2x + 5)$

2. Factorise the following, where possible: (a) $3x^2 - 7x$, (b) $3x + 6x^2$, (c) $x^2 + 3x + 2$, (d) $x^2 - 4$, (e) $x^2 - x - 2$, (f) $2x^2 + 5$, (g) $3x^2 + 8x + 5$, (h) $x^2 - 8x + 7$, (i) $4x^2 - 25$, (j) $6x^2 + 7x - 5$, (k) $2x^2 + 7x - 2$, (l) $4x^2 - 12x + 9$

15.2 SOLUTION OF QUADRATIC EQUATIONS BY FACTORISATION

The solutions of a quadratic equation are the values of x which make the quadratic expression equal to zero. As was seen by the graphs of chapter 3 a quadratic equation normally has 2 solutions. The quadratic equation has two solutions or no solutions. When the equation appears to have no solutions it is because the solutions are not real numbers, they are complex numbers. Equations with no real solutions will be considered in later years.
When the solutions are whole numbers or simple fractions they can be found by factorising the quadratic expression. When a quadratic expression is factorised it has two factors multiplied together. In a quadratic equation the product of these factors is zero and hence one of the factors must be zero.

Example 15.2 Solve the equation $x^2 - 3x + 2 = 0$

$x^2 - 3x + 2$ can be factorised to $(x - 2)(x - 1)$

Hence $(x - 2)(x - 1) = 0$ and therefore $x - 2 = 0$ or $x - 1 = 0$

When $x - 2 = 0$, $x = 2$

When $x - 1 = 0$, $x = 1$

These solutions can be checked in the original equation.

When $x = 1; x^2 - 3x + 2 = 1^2 - 3 \times 1 + 2 = 0$

When $x = 2; x^2 - 3x + 2 = 2^2 - 3 \times 2 + 2 = 0$

Hence $x = 1$ and $x = 2$ are the solutions of $x^2 - 3x + 2 = 0$

Example 15.3 Solve the following quadratic equations:

(a) $3x^2 - 17x + 10 = 0$, (b) $8x^2 - 2x = 0$,

(c) $4x^2 - 9 = 0$, (d) $x^2 - 6x + 9 = 0$

(a) $3x^2 - 17x + 10 = 3x^2 - 2x - 15x + 10$
$$= x(3x - 2) - 5(3x - 2)$$
$$= (3x - 2)(x - 5)$$

Therefore $(3x - 2)(x - 5) = 0$

Either $3x - 2 = 0$, $3x = 2$, $x = \frac{2}{3}$

or $x - 5 = 0$, $x = 5$

Check: $x = \frac{2}{3}; 3(\frac{2}{3})^2 - 17(\frac{2}{3}) + 10 = \frac{4}{3} - \frac{34}{3} + 10 = 0$

$x = 5; 3(5)^2 - 17(5) + 10 = 75 - 85 + 10 = 0$

Solutions are $x = \frac{2}{3}$ or $x = 5$

(b) $8x^2 - 2x = 2x(4x - 1)$ and $2x(4x - 1) = 0$

Therefore
either $2x = 0$, $x = 0$

or $4x - 1 = 0$, $4x = 1$, $x = \frac{1}{4}$

Check: $x = 0; 8(0)^2 - 2(0) = 0$

$x = \frac{1}{4}; 8(\frac{1}{4})^2 - 2(\frac{1}{4}) = \frac{1}{2} - \frac{1}{2} = 0$

Solutions are $x = 0$ or $x = \frac{1}{4}$

The solution $x = 0$ must not be omitted, it is a solution in the same way as $x = \frac{1}{4}$ is a solution. Equations should not be divided throughout by variables, such as x, since this removes one of the solutions of the equation.

(c) In this example there is no need to factorise, the factors are $(2x - 3)(2x + 3)$ but it is usual to work as follows.

$4x^2 - 9 = 0$, add 9 to both sides

$4x^2 = 9$, divide both sides by 4

$x^2 = \frac{9}{4}$, take the square root of both sides

$x = \pm\frac{3}{2}$, remember to include the $\pm$ since there are two square roots of $\frac{9}{4}$; $+\frac{3}{2}, -\frac{3}{2}$

$x = \frac{3}{2}$ or $x = -\frac{3}{2}$

Check: $x = \frac{3}{2}, 4(\frac{3}{2})^2 - 9 = 9 - 9 = 0$

$x = -\frac{3}{2}, 4(-\frac{3}{2})^2 - 9 = 9 - 9 = 0$

Solutions are $x = 1\frac{1}{2}$ or $x = -1\frac{1}{2}$

(d) $x^2 - 6x + 9 = (x - 3)(x - 3)$
$$(x - 3)(x - 3) = 0$$

There is only one solution to this equation, or we say there are two solutions but they are the same number. This occurs when the quadratic expression is a perfect square. If the graph were drawn it would touch the x-axis instead of cutting it in two places.

Check: $x = 3; (3)^2 - 6(3) + 9 = 9 - 18 + 9 = 0$

Solution is $x = 3$

EXERCISE 15.2

Solve the following quadratic equations

1. $x^2 + 2x - 15 = 0$, 2. $6x^2 - 8x = 0$, 3. $6x^2 - 37x + 56 = 0$,
4. $x^2 - 16 = 0$, 5. $3x^2 - 23x + 14 = 0$, 6. $4x^2 - 12x + 9 = 0$,
7. $9x^2 - 4 = 0$, 8. $2x^2 + 13x + 15 = 0$

15.3 SOLUTION OF QUADRATIC EQUATIONS BY FORMULA

In general quadratic equations do not factorise and therefore some other method of solving quadratic equations must be used. A formula can be derived to solve quadratic equations.

Consider the equation $ax^2 + bx + c = 0$
$$ax^2 + bx = -c$$

Multiply both sides by $4a$ and then add b^2 to both sides
$$4a^2x^2 + 4abx + b^2 = -4ac + b^2$$

The left hand side is a perfect square and can be factorised
$$(2ax + b)(2ax + b) = b^2 - 4ac$$

Take the square root of both sides
$$2ax + b = \pm\sqrt{b^2 - 4ac}$$

Subtract b from both sides
$$2ax = -b \pm \sqrt{b^2 - 4ac}$$

Divide both sides by $2a$
$$x = \frac{-b \pm \sqrt{b^2 - 4ac}}{2a}$$

This formula can be applied to any quadratic equation. The important part of this formula is $b^2 - 4ac$
$b^2 - 4ac$ is called the discriminant.
When $b^2 - 4ac$ is positive, the square root can be found, and the equation has two real roots.
When $b^2 - 4ac$ is zero, the equation has two equal roots.
When $b^2 - 4ac$ is negative, the equation has no real roots since there is no real square root of a negative number.

Example 15.4 Solve the equations:

(a) $10x^2 + 4x - 3 = 0$, (b) $x^2 - 5.486x + 7.524 = 0$,

(c) $\dfrac{7}{2x + 3} - \dfrac{9}{x + 8} = 1$

(a) $10x^2 + 4x - 3 = 0$

$$a = 10, b = 4, c = -3$$

Using the above formula:

$$x = \frac{-4 \pm \sqrt{4^2 - 4 \times 10 \times -3}}{2 \times 10}$$

$$x = \frac{-4 \pm \sqrt{136}}{20}$$

$$x = \frac{-4 \pm 11.66}{20}$$

$$x = \frac{-4 + 11.66}{20} = 0.3831$$

$$x = \frac{-4 - 11.66}{20} = -0.7830$$

Check: $x = 0.3831$; $10(0.3831)^2 + 4(0.3831) - 3$

$$= 1.4677 + 1.5324 - 3 = 0.0001$$

$x = -0.7830$; $10(-0.7830)^2 + 4(-0.7830) - 3$

$$= 6.1309 - 3.1320 - 3 = -0.0011$$

When the answers are decimals the check will not always come to zero. This is because the answers have been rounded off.

(b) $x^2 - 5.486x + 7.524 = 0$

$$a = 1, b = -5.486, c = 7.524$$

Using the formula:

$$x = \frac{-(-5.486) \pm \sqrt{(-5.486)^2 - 4 \times (1) \times (7.524)}}{2 \times 1}$$

$$x = \frac{5.486 \pm \sqrt{30.10 - 30.10}}{2} = \frac{5.486}{2} = 2.743$$

Check: $x = 2.743$, $(2.743)^2 - 5.486 \times (2.743) + 7.524$

$$= 7.524 - 15.048 + 7.524 = 0$$

This quadratic expression is a perfect square when working to 4 significant figures, and hence there is only one solution $x = 2.743$

(c) $\dfrac{7}{2x + 3} - \dfrac{9}{x + 8} = 1$

We must first clear the fractions by multiplying by $(2x + 3)$ and by $(x + 8)$

$$(2x + 3) \times (x + 8) \times \frac{7}{2x + 3} - (2x + 3)(x + 8) \times \frac{9}{x + 8}$$

$$= 1 \times (2x + 3) \times (x + 8)$$

$$7(x + 8) - 9(2x + 3) = (2x + 3)(x + 8)$$

$$7x + 56 - 18x - 27 = 2x^2 + 16x + 3x + 24$$

$$-11x + 29 = 2x^2 + 19x + 24$$

Subtract $2x^2 + 19x + 24$ from both sides

$$-11x + 29 - (2x^2 + 19x + 24) = 0$$

$$-11x + 29 - 2x^2 - 19x - 24 = 0$$

$$-2x^2 - 30x + 5 = 0$$

It is usual to make the x^2 term positive by multiplying both sides by -1

$$2x^2 + 30x - 5 = 0$$

$$a = 2, b = 30, c = -5$$

Using the formula

$$x = \frac{130 \pm \sqrt{30^2 - 4 \times 2 \times (-5)}}{2 \times 2}$$

$$x = \frac{-30 \pm \sqrt{940}}{4} = \frac{-30 \pm 30.66}{4}$$

$$x = \frac{-30 + 30.66}{4} = \frac{0.66}{4} = 0.165$$

$$x = \frac{-30 - 30.66}{4} = \frac{-60.66}{4} = -15.165$$

Check: $x = 0.165$; $\dfrac{7}{2(0.165) + 3} - \dfrac{9}{0.165 + 8}$

$$= \frac{7}{3.33} - \frac{9}{8.165}$$

$$= 2.102 - 1.102 = 1.000$$

$x = -15.165$; $\dfrac{7}{2(-15.165) + 3} - \dfrac{9}{-15.165 + 8}$

$$= \frac{7}{-27.33} - \frac{9}{-7.165}$$

$$= -0.256 + 1.256 = 1.000$$

Solutions are $x = 0.165$ or $x = -15.165$

EXERCISE 15.3

Solve the following quadratic equations

1. $x^2 - 2x - 1 = 0$, 2. $3x^2 - 7x + 1 = 0$, 3. $9x^2 - 5 = 0$,

4. $x^2 + 5x + 3 = 0$, 5. $3x^2 + x - 5 = 0$,

6. $2.5x^2 - 4.2x - 1.5 = 0$, 7. $\dfrac{5}{x - 1} + \dfrac{12}{x + 5} = 3$,

8. $\dfrac{2x}{5-x} + \dfrac{3}{x+1} = 8$

15.4 PROBLEMS INVOLVING QUADRATIC EQUATIONS

Example 15.5 The height in metres, after t seconds, of a body projected vertically, is given by $h = 35t - 5t^2$. Find the times when the body is 12 m above its point of projection.

When $h = 12$ m $\qquad 12 = 32t - 5t^2$

Subtract $32t - 5t^2$ from both sides:

$$12 - (32t - 5t^2) = 0$$
$$12 - 32t + 5t^2 = 0$$
$$5t^2 - 32t + 12 = 0$$
$$5t^2 - 32t + 12 = 5t^2 - 2t - 30t + 12$$
$$= t(5t - 2) - 6(5t - 2)$$
$$= (t - 6)(5t - 2)$$
$$(t - 6)(5t - 2) = 0$$

Either $t - 6 = 0, t = 6$

or $5t - 2 = 0, 5t = 2, t = 0.4$

Check: $t = 6, h = 32 \times 6 - 5 \times 6^2 = 192 - 180 = 12$

$t = 0.4, h = 32 \times 0.4 - 5 \times 0.4^2 = 12.8 - 0.8 = 12$

The body is 12 m from its point of projection at 0.4 s and 6 s.

Example 15.6 A piece of sheet metal consists of a rectangle of length 800 mm and width $2r$ mm with a semi-circle of radius r mm on one end. If the area of the metal is 720 000 mm^2, form an equation and hence determine the value of r.

Area of semi-circle + area of rectangle = total area

$$\frac{\pi r^2}{2} + 800 \times 2r = 720\,000$$

$$1.571r^2 + 1600r = 720\,000$$

Subtract 720 000 from both sides

$$1.571r^2 + 1600r - 720\,000 = 0$$

$$a = 1.571, b = 1600, c = -720\,000$$

Using the formula

$$r = \frac{-1600 \pm \sqrt{1600^2 - 4 \times 1.571 \times (-720\,000)}}{2 \times 1.571}$$

$$r = \frac{-1600 \pm \sqrt{7\,084\,000}}{3.142}$$

$$r = \frac{-1600 \pm 2662}{3.142}$$

$$r = \frac{-1600 + 2662}{3.142} = \frac{1062}{3.142} = 338$$

$$r = \frac{-1600 - 2662}{3.142} = \frac{-4262}{3.142} = -1356$$

In problems involving quadratic equations both roots are not always solutions to the given problem. Sometimes a check is made to see which is the correct value for the solution. In this case a check is not necessary since it is obvious that r must be positive

Check: $x = 338$; Area of rectangle $= 800 \times (2 \times 338)$

$$= 540\,800$$

$$\text{Area of semi-circle} = \frac{3.142 \times 338^2}{2} = 179\,500$$

Total area $= 540\,800 + 179\,500 = 720\,300$

Radius $= 338$ mm

EXERCISE 15.4

1. A closed cylindrical can has a total surface area of 48 000 mm^2. Find the radius of its circular ends if its height is 100 mm.

2. The sum, S, of a number of terms, n, of an arithmetic progression is given by $S = \dfrac{n}{2}(2a + [n-1]d)$ where a is the first term, and d is the common difference. Find how many terms n need to be added to make the sum $S = 990$ if $a = 2$ and $d = 5$.

3. A uniform metal disc, radius r, is to have a flat of width w milled on its curved surface. The maximum depth of the segment removed is h. Show that, r, w, and h are connected by the equation $h^2 - 2rh + \dfrac{w^2}{4} = 0$.

Calculate the volume of metal removed when $r = 80$ mm, $w = 40$ mm and the thickness of the disc is 5 mm.

4. The relationship between the grid voltage V volts and the anode current I mA of a valve in a square law modulator is $I = V^2 + 17.2V + 68.61$.
At what grid voltage will the valve cut off, (i.e. I becomes zero).

5. Fig. 15.1 shows an unequal right-angled section. The width of each leg x mm is to be the same and the cross-sectional area of 5000 mm^2. Claculate the width of each leg.

6. Fig. 15.2 shows a standard metal blank. Given that the area is 23 100 mm^2 and the length of the rectangle is 110 mm form a quadratic equation and solve it to find the radius R.

7. Transpose the expression $W = 2\sqrt{Dh - h^2}$ to form the quadratic equation $4h^2 - 4Dh + W^2 = 0$.
Hence find the values of h when $W = 4$ and $D = 5$.

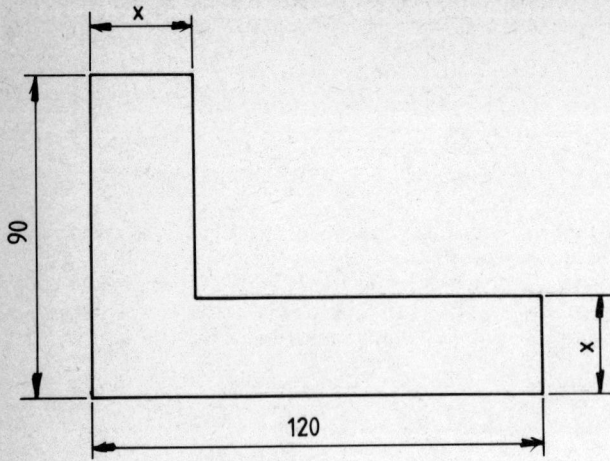

Figure 15.1

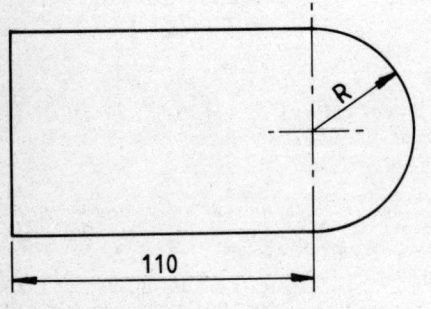

Figure 15.2

15.5 SIMULTANEOUS EQUATIONS, ONE LINEAR ONE QUADRATIC

Use the linear equation to find one unknown in terms of the other. Substitute this into the quadratic equation.

Example 15.7 Solve the simultaneous equations

$2x - y = 1$ (1) $x^2 + 3xy + y^2 = 31$ (2)

From (1) $y = 2x - 1$; substitute this into (2)

$$x^2 + 3x(2x - 1) + (2x - 1)^2 = 31$$
$$x^2 + 6x^2 - 3x + 4x^2 - 4x + 1 = 31$$
$$11x^2 - 7x + 1 = 31$$
$$11x^2 - 7x - 30 = 0$$
$$11x^2 - 7x - 30 = 11x^2 + 15x - 22x - 30$$
$$= x(11x + 15) - 2(11x + 15)$$
$$= (x - 2)(11x + 15)$$
$$(x - 2)(11x + 15) = 0$$

Either $x - 2 = 0, x = 2$

or $11x + 15 = 0, 11x = -15, x = -\dfrac{15}{11} = -1.364$

From (1); when $x = 2, 4 - y = 1, y = 3$

when $x = -1.364, -2.728 - y = 1,$

$$y = -3.728$$

Check in (2) $x = 2, y = 3,$

$$x^2 + 3xy + y^2 = 2^2 + 3 \times 2 \times 3 + 3^2$$

$$= 4 + 18 + 9 = 31$$

$x = -1.364, y = -3.728,$

$$x^2 + 3xy + y^2 = (-1.364)^2 + 3 \times (-1.364)$$
$$\times (-3.728) + (-3.728)^2$$
$$= 1.8605 + 15.2550 + 13.898$$
$$= 31.01$$

Solutions are $x = 2, y = 3$ and $x = -1.364,$ $y = -3.728$

EXERCISE 15.5

Solve the following simultaneous equations

1. $x - 2y = 1$ 2. $3x + 2y = 8$

   $x^2 - 3xy + 2y^2 = 4$ $3x^2 - y^2 = 11$

3. $x - y = 5$ 4. $2x + 3y = 8$

   $2xy - y^2 = 56$ $x^2 - xy + y^2 = 3$

15.6 GENERAL EXAMPLES

Example 15.8 Fig. 15.3 shows the diagram of a steel framework which is symmetrical about the vertical member AB. From the measurements given in the figure calculate the lengths of the members AC and AB, all measurements are in metres.

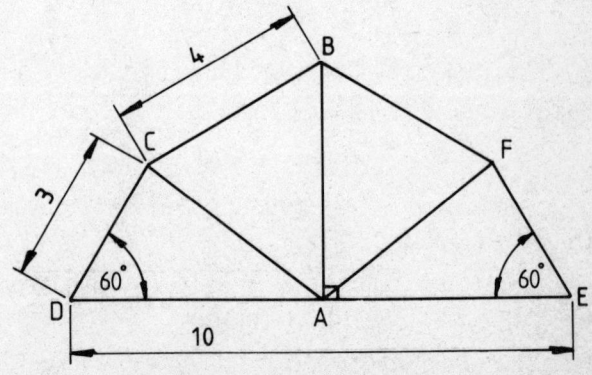

Figure 15.3

In $\triangle$ACD, DA = 5 m, DC = 3 m, $\angle$CDA = $60°$

Using the cosine rule

$$AC^2 = 3^2 + 5^2 - 2 \times 3 \times 5 \times \cos 60°$$

$$AC^2 = 9 + 25 - 15 = 19$$

$$AC = \sqrt{19} = 4.36 \text{ m}$$

Using the sine rule

$$\frac{3}{\sin \angle CAD} = \frac{4.36}{\sin 60°}$$

$$\sin \angle CAD = \frac{3 \times \sin 60°}{4.36} = 0.5959$$

$$\angle CAD = 36.58°$$

$$\angle BAC = 90° - \angle CAD = 90° - 36.58°$$

$$\angle BAC = 53.42°$$

In $\triangle$ABC, CB = 4 m, AC = 4.36 m, $\angle$BAC = $53.42°$

Using the cosine rule

$$BC^2 = AC^2 + AB^2 - 2 \times AC \times AB \times \cos \angle BAC$$

$$16 = 19 + AB^2 - 2 \times 4.36 \times AB \times 0.5959$$

$$16 = 19 + AB^2 - 5.196AB$$

$$AB^2 - 5.196AB + 3 = 0$$

Using the formula $a = 1$, $b = -5.196$, $c = 3$

$$AB = \frac{5.196 \pm \sqrt{(-5.196)^2 - 4 \times 1 \times 3}}{2 \times 1}$$

$$AB = \frac{5.196 \pm \sqrt{15}}{2} = \frac{5.196 \pm 3.873}{2}$$

$$AB = 4.53 \text{ m or } 0.662 \text{ m}$$

By inspection of the diagram it can be seen that AB = 4.53 m. Hence the length of members are AC = 4.36 m, AB = 4.53 m.

EXERCISE 15.6

1. The weight of a framed structure is determined by the formula $W = \frac{3dL}{4}\left(1 + \frac{L}{10}\right)$ Calculate the value of the span L when $W = 1500$ and $d = 10$.

2. A rectangular hole is cut out from a 250 mm $\times$ 200 mm uniform plate. The distance x between the edge of the plate and the edge of the hole is the same on all sides. The hole reduces the weight by 60%.
(a) Show that $x^2 - 225x + 5000 = 0$
(b) Solve this equation for x and hence determine the dimensions of the hole.

3. The Weston cell is used as a standard e.m.f. of 1.0183 V at $20°$C. The e.m.f. varies so that when the temperature is $(20 + x)°$C the correction is $-(x^2 + 41x) \times 10^{-6}$ V. Calculate the temperature at which a correction of $-0.000\,560$ V is required assuming the temperature is greater than $20°$C.

4. A vertical storage tank has a cylindrical body, a hemispherical top, and a flat bottom. The cylindrical part of the body is of external diameter d and height h, and the external radius of the top is $\frac{d}{2}$. Show that the total external surface area, A, of the tank is given by

$$A = 3\pi\left(\frac{d^2}{4} + \frac{dh}{3}\right)$$

Find a suitable value for d when $A = 24\pi$ m^2 and $h = 3$ m.

5. In fig. 15.3, ED = 8 m, DC = 2 m, CB = 4 m and $\angle$CDA = $65°$. Calculate the lengths of AC and AB.

SUMMARY

1. When a quadratic equation has whole number or simple fractions as solutions it can be solved by first finding the factors of the quadratic expression.

2. When simplifying a quadratic equation it must not be divided through by a variable or one of the solutions to the equation will be lost.

3. When a quadratic equation does not factorise it can be solved by using the formula. If $ax^2 + bx + c = 0$

$$x = \frac{-b \pm \sqrt{b^2 - 4ac}}{2a}$$

4. When $b^2 - 4ac$ is negative there are no real solutions to the equation.

5. When $b^2 - 4ac = 0$ there is only one solution to the equation (or two identical solutions).

6. When a problem is solved by a quadratic equation both of the solutions to the equation may not be solutions to the problem. The solutions should always be checked in the problem to see that they are solutions to the given problem.

7. When a pair of simultaneous equations are such that one is a linear equation and one a quadratic then these are solved by substituting one unknown in terms of the other from the linear equation into the quadratic.

SELF ASSESSMENT PAPER No 15

Instructions: Answer all the questions in both sections
Time allowed: Section A 30 minutes (35 marks)
   Section B 1 hour
   (25 marks each question)
Marks gained: 55+ pass with credit, 44—55 pass, less than 44 fail, repeat chapter 15

Section A

1. Factorise the following (a) $x^2 - 3x$, (b) $9x^2 - 4$, (c) $x^2 - 3x + 2$, (d) $3x^2 + 11x + 10$

2. Solve the equations (a) $x^2 - 6x + 5 = 0$, (b) $4x^2 - 25 = 0$, (c) $3x^2 = 5x$

Section B

1. Solve the following equations when the equations have real solutions. Otherwise state that no solution is possible.

(a) $2x^2 - 5x - 3 = 0$, (b) $2x^2 + 5x + 4 = 0$

(c) $x^2 - 10.4x + 27.04 = 0$

2. A sheet metal air duct is to have a rectangular cross-section such that its width is 60 mm less than its length. If the total cross-sectional area is to be 56 000 mm^2, find the dimensions of the duct.

3. Solve the simultaneous equations $2x + y = 7$

$$x^2 - xy = 6$$

ANSWERS

Exercise 15.1 1. (a) $2x^2 - 6x$, (b) $2x^2 + 9x - 5$,
(c) $5x^2 - 33x - 14$, (d) $9x^2 + 24x + 16$, (e) $9x^2 - 25$,
(f) $14x^2 + 29x - 15$, 2. (a) $x(3x - 7)$,
(b) $3x(1 + 2x)$, (c) $(x + 2)(x + 1)$, (d) $(x - 2)(x + 2)$,
(e) $(x - 2)(x + 1)$, (f) no factors, (g) $(3x + 5)(x + 1)$,
(h) $(x - 7)(x - 1)$ (i) $(2x - 5)(2x + 5)$,
(j) $(2x - 1)(3x + 5)$, (k) no factors, (l) $(2x - 3)^2$

Exercise 15.2 1. $3, - 5$, 2. $0, 1\frac{1}{3}$,
3. $3\frac{1}{2}, 2\frac{2}{3}$, 4. $- 4, 4$, 5. $\frac{2}{3}, 7$,
6. $1\frac{1}{2}$, 7. $\pm \frac{2}{3}$, 8. $- 5, -\frac{3}{2}$

Exercise 15.3 1. $2.414, - 0.414$, 2. 2.181,
0.153, 3. ± 0.7454, 4. $- 4.303, - 0.697$,
5. $1.135, - 1.468$, 6. $1.983, - 0.303$,
7. $4, - 2\frac{1}{3}$, 8. $3.936, - 0.636$.

Exercise 15.4 1. 50.7 mm, 2. 20,
3. 374 mm^3, 4. $- 6.3$ V, $- 10.9$ V,
5. 27.4 mm, 6. 70 mm, 7. $1, 4$

Exercise 15.5 1. $x = 7, y = 3$, 2. $x = 2$,
$y = 1; x = - 18, y = 31$, 3. $x = 9, y = 4$,
$x = - 9, y = - 14$, 4. $x = 1, y = 2$;
$x = 1.947\left(\dfrac{37}{19}\right), y = 1.368\left(\dfrac{26}{19}\right)$

Exercise 15.6 1. 40, 2. 200 mm $\times$ 150 mm,
3. $30.8°$, 4. 2.0 m, 5. $AC = 3.64$ m,
$AB = 4.27$ m.

SELF ASSESSMENT PAPER No 15

	Marks
Section A	
1. (a) $x(x - 3)$	3
(b) $(3x - 2)(3x + 2)$	4
(c) $(x - 2)(x - 1)$	4
(d) $(3x + 5)(x + 2)$	4
2. (a)$(x - 5)(x - 1)$	4
$x = 5$ or $x = 1$	2

	Marks
2. (b) $4x^2 = 25$	2
$x^2 = \dfrac{25}{4}$	2
$x = \pm\frac{5}{2}$	2
(c) $3x^2 - 5x = 0$	2
$x(3x - 5) = 0$	4
$x = 0$ or $x = 1\frac{2}{3}$	2

Section B

1. (a) $a = 2, b = - 5, c = - 3$	1
$x = \dfrac{-(- 5) \pm \sqrt{(- 5)^2 - 4 \times 2 \times (- 3)}}{2 \times 2}$	4
$x = \dfrac{5 \pm \sqrt{49}}{4}$	2
$x = 3$ or $x = -\frac{1}{2}$	2
(b) $a = 2, b = 5, c = 4$	1
$x = \dfrac{- 5 \pm \sqrt{5^2 - 4 \times 2 \times 4}}{2 \times 2}$	4
$x = \dfrac{- 5 \pm \sqrt{- 7}}{4}$	2
No solution possible	1
(c) $a = 1, b = - 10.4, c = 27.04$	1
$x = \dfrac{-(- 10.4) \pm \sqrt{(- 10.4)^2 - 4 \times 1 \times 27.04}}{2 \times 1}$	4
$x = \dfrac{10.4 \pm \sqrt{0}}{2}$	2
$x = 5.2$	1

2. Let the length be x mm	1
Area of cross-section $= x(x - 60)$	4
$x(x - 60) = 56\,000$	4
$x^2 - 60x - 56\,000 = 0$	2
$a = 1, b = - 60, c = - 56\,000$	1
$x = \dfrac{-(- 60) \pm \sqrt{(- 60)^2 - 4 \times 1 \times (- 56\,000)}}{2 \times 1}$	4
$x = \dfrac{60 \pm \sqrt{227\,600}}{2}$	2
$x = \dfrac{60 \pm 477.1}{2}$	2
$x = 269$ mm other solution is negative	3
Length $= 269$ mm, width $= 209$ mm	2

3. $y = 7 - 2x$	2
$x^2 - x(7 - 2x) = 6$	4
$x^2 - 7x + 2x^2 = 6$	4
$3x^2 - 7x - 6 = 0$	2

Marks

$a = 3, b = -7, c = -6$

1

$$x = \frac{-(-7) \pm \sqrt{(-7)^2 - 4 \times 3 \times (-6)}}{2 \times 3}$$

4

Marks

$$x = \frac{7 \pm \sqrt{121}}{6}$$

2

$x = 3$ or $x = -\frac{2}{3}$

3

$y = 1$ or $y = 8\frac{1}{3}$

3

16 Indices and Logarithms

16.1 LAWS OF INDICES

For any values of m and n, positive, negative or fractional

(1) $a^m \times a^n = a^{m+n}$

(2) $a^m \div a^n = a^{m-n}$

(3) $(a^m)^n = a^{mn}$

(4) If in (3) m is replaced by $\dfrac{1}{n}$ then $(a^{\frac{1}{n}})^n = a^{\frac{1}{n} \times n} = a^1$

$$\text{or } a^{\frac{1}{n}} = \sqrt[n]{a}$$

(5) If (4) is raised to the power m then

$$(a^{\frac{1}{n}})^m = (\sqrt[n]{a})^m$$

$$a^{\frac{m}{n}} = (\sqrt[n]{a})^m = \sqrt[n]{a^m}$$

(6) If $m = 0$ in (1) then $a^0 \times a^n = a^{0+n} = a^n$

$$\text{Hence } a^0 = 1$$

(7) If $m = -n$ in (1) then $a^{-n} \times a^n = a^{-n+n} = a^0 = 1$

$$\text{Hence } a^{-n} = \frac{1}{a^n}$$

Example 16.1 Find the values of (a) $9^{\frac{3}{2}}$, (b) $5^{-\frac{1}{2}}$, (c) $(8^{\frac{1}{3}})^4$,

(a) $9^{\frac{3}{2}} = \sqrt{9^3} = (\sqrt{9})^3 = 3^3 = 27$

(b) $5^{-\frac{1}{2}} = \dfrac{1}{5^{\frac{1}{2}}} = \dfrac{1}{\sqrt{5}} = \dfrac{1}{2.236} = 0.4472$

(c) $(8^{\frac{1}{3}})^4 = (\sqrt[3]{8})^4 = 2^4 = 16$

EXERCISE 16.1

Evaluate

1. (a) $16^{\frac{3}{4}}$, (b) $27^{\frac{2}{3}}$, (c) $64^{\frac{1}{6}}$

2. (a) 5^{-2}, (b) $27^{-\frac{2}{3}}$, (c) $8^{-\frac{5}{3}}$, (d) $32^{-\frac{1}{5}}$

3. (a) $(7^{\frac{1}{3}})^2$, (b) $5^{-\frac{1}{4}}$, (c) $3^{\frac{3}{2}}$,   4. $(2^3 \times 2^{-1}) \div 2^2$

5. $3^{-1}(3^3 + 3^2)$,   6. $(2^{-1})^{-2} \times (2^{-\frac{1}{2}})^2$

16.2 LAWS OF LOGARITHMS

If $y = a^x$ then x is the logarithm of y to base a and this is written $x = \log_a y$

The logarithm is an index and therefore logarithms should obey the same laws as indices. Hence if we look at the laws of indices in 16.1 we have:

(1) if $x = a^m$ and $y = a^n$ then $xy = a^{m+n}$
From the definition for logarithms $\log_a xy = m + n$ also $\log_a x = m$ and $\log_a y = n$

$$\text{Hence } \log_a xy = \log_a x + \log_a y$$

(2) Similarly if $x = a^m$, $y = a^n$ then $x \div y = a^m \div a^n = a^{m-n}$

$$\log_a(x \div y) = m - n = \log_a x - \log_a y$$

(3) If $x = a^m$ then $x^n = (a^m)^n = a^{mn}$

$$\log_a x^n = mn = n \times \log_a x$$

$$\log_a x^n = n \log_a x$$

In particular:

If $n = \dfrac{1}{q}$, $\log_a \sqrt[q]{x} = \log_a x^{\frac{1}{q}} = \dfrac{1}{q} \log_a x$

If $n = -q$, $\log_a x^{-q} = -q \log_a x$

Also $\log_a a = 1$ ($y = a^x$ then $x = \log_a y$, when $y = a$, $x = 1$)

$$\log_a 1 = 0 \ (y = a^x \text{ then } x = \log_a y, \text{ when } y = 1, x = 0)$$

$$\log_a \frac{1}{x} = -\log_a x \left(\log_a \frac{1}{x} = \log_a 1 - \log_a x = 0 - \log_a x\right)$$

Before electronic calculators were available common logarithms to base 10 were used as an aid to calculating. Now the logarithm tables to base 10 are very seldom used. The laws of logarithms are, however, very important and will prove useful in future work. Here we will look at logarithms to any base.

Example 16.2 (a) Reduce the following to a single logarithm
(i) $\log 6 - 2 \log 3 + \log 5$, (ii) $2 \log 2 - \log 8 + \log 12$
(b) Solve the equation $2 \log 2 - \log 6 + 2 \log 3 = \log x$

(a) (i) $\log 6 - 2\log 3 + \log 5 = \log 6 - \log 3^2 + \log 5$
$$= \log 6 - \log 9 + \log 5$$
$$= \log 6 \div 9 \times 5$$
$$= \log \frac{6 \times 5}{9} = \log \frac{10}{3}$$

(ii) $2\log 2 - \log 8 + \log 12 = \log 2^2 - \log 8 + \log 12$
$$= \log 4 - \log 8 + \log 12$$
$$= \log \frac{4 \times 12}{8} = \log 6$$

(b) $2\log 2 - \log 6 + 2\log 3 = \log x$
$$\log 2^2 - \log 6 + \log 3^2 = \log x$$
$$\log 4 - \log 6 + \log 9 = \log x$$
$$\log \frac{4 \times 9}{6} = \log x$$
$$\log 6 = \log x \quad \text{hence } x = 6$$

Example 16.3 (a) Find the value of $\log_4 64$, (b) Find the value of a when $\log_a 32 = 5$

(a) $\log_4 64 = \log_4 4^3 = 3\log_4 4 = 3 \ (\log_a a = 1)$

(b) $\log_a 32 = \log_a 2^5 = 5\log_a 2$

Hence $5\log_a 2 = 5$ or $\log_a 2 = 1$ or $a = 2$

This could also be solved as follows:

$\log_a 32 = 5$, hence $32 = a^5$, $2^5 = a^5 \quad a = 2$

EXERCISE 16.2

Reduce the following to single logarithms:

1. $\log 2 - 3\log 2 + \log 3$,   2. $3\log 3 - \log 6 + \log 2$

3. $\log 18 - 3\log 3 + 2\log 3$, 4. $2\log 2 - 3\log 4 + \log 6$

5. Find the values of (a) $\log_3 27$, (b) $2\log_4 2$,

6. Find the value of a when (a) $\log_a 81 = 4$, (b) $\log_a 125 = 3$

7. Evaluate $\dfrac{\log 16 - \log 2}{\log 4}$

8. Solve the following equations for x:

(a) $\log 3 + \log 8 - \log 12 = \log x$

(b) $4\log 2 + \log 4 - \log 8 = x\log 2$

16.3 INDICIAL EQUATIONS

Certain equations, in which the unknown x occurs in the index can be solved by taking logarithms of both sides of the equation. It must be remembered that, in taking logarithms, it is the logarithm of **all** of the left hand side which is equal to the logarithm of **all** of the right hand side and that $\log (a + b)$ does not equal $\log a + \log b$.

Example 16.4 Solve the equations (a) $2^x = 7$, (b) $3^x . 5^{x-1} = 4^{x+1}$

(a) $\log_{10} 2^x = \log_{10} 7$
$$x\log_{10} 2 = \log_{10} 7$$

Either from tables or from a calculator $\log_{10} 2 = 0.3010$
$$\log_{10} 7 = 0.8451$$
$$x = \frac{\log_{10} 7}{\log_{10} 2} = \frac{0.8451}{0.3010} = 2.808$$
$$x = 2.808$$

(b) $\log_{10}(3^x . 5^{x-1}) = \log_{10} 4^{x+1}$
$$\log_{10} 3^x + \log_{10} 5^{x-1} = (x + 1)\log_{10} 4$$
$$x\log_{10} 3 + (x - 1)\log_{10} 5 = (x + 1)\log_{10} 4$$

Either from tables or from a calculator $\log_{10} 3 = 0.4771$
$$\log_{10} 5 = 0.6990$$
$$\log_{10} 4 = 0.6021$$
$$0.4771x + 0.6990(x - 1) = 0.6021(x + 1)$$
$$0.4771x + 0.6990x - 0.6990 = 0.6021x + 0.6021$$
$$(0.4771 + 0.6990 - 0.6021)x = 0.6021 + 0.6990$$
$$0.5740x = 1.3011$$
$$x = \frac{1.3011}{0.5740} = 2.267$$
$$x = 2.267$$

EXERCISE 16.3

1. Solve for x. (a) $3^x = 17$, (b) $6^x = 2.5^{x+1}$, (c) $5^x . 4^{x-1} = 7^{x+1}$ (d) $x^{3.25} = 73$

2. Solve the equations: (a) $(2x)^{2.5} = 50$, (b) $(2.5)^{2x} = 50$

3. Solve for x: $7^x = 3^{x+1} . 2^{x-2}$

4. Solve for x: $5^x = (4.5)^{x+1} . 7^{x-1}$

16.4 GENERAL EXAMPLES

Example 16.5 (a) Solve the equation $3^{x^2+1} = 9^{2x}$

(b) Solve the equation $25^x - 5^{x+1} + 6 = 0$

(a) $3^{x^2+1} = 9^{2x}$ The 9 can be written as 3^2
$$3^{x^2+1} = (3^2)^{2x} \quad \text{or} \quad 3^{x^2+1} = 3^{4x}$$
$$\text{Hence } x^2 + 1 = 4x$$
$$x^2 - 4x + 1 = 0$$

Using the formula for a quadratic equation
$$a = 1, b = -4, c = 1$$
$$x = \frac{-b \pm \sqrt{b^2 - 4ac}}{2a} = \frac{4 \pm \sqrt{(-4)^2 - 4 \times 1 \times 1}}{2 \times 1}$$

$$x = \frac{4 \pm \sqrt{12}}{2} = \frac{4 \pm 3.464}{2}$$

$$x = \frac{4 + 3.464}{2} = 3.732$$

$$\text{or } x = \frac{4 - 3.464}{2} = 0.268$$

Solutions are $x = 0.268$ or $x = 3.732$

(b) $25^x - 5^{x+1} + 6 = 0$

The 25^x can be written $(5^2)^x = 5^{2x} = (5^x)^2$

and the $5^{x+1} = 5^x.5 = 5(5^x)$

The equation becomes $(5^x)^2 - 5(5^x) + 6 = 0$

put $y = 5^x$

$$y^2 - 5y + 6 = 0$$
$$(y - 3)(y - 2) = 0$$
$$y = 3 \text{ or } y = 2$$

When $y = 3$, $5^x = 3$ take logarithms of both sides

$$\log_{10} 5^x = \log_{10} 3$$
$$x \log_{10} 5 = \log_{10} 3$$
$$x = \frac{\log_{10} 3}{\log_{10} 5} = \frac{0.4771}{0.6990} = 0.683$$

When $y = 3$, $5^x = 2$ take logarithms of both sides

$$\log_{10} 5^x = \log_{10} 2$$
$$x \log_{10} 5 = \log_{10} 2$$
$$x = \frac{\log_{10} 2}{\log_{10} 5} = \frac{0.3010}{0.6990} = 0.431$$

The solutions are $x = 0.431$ or $x = 0.683$

EXERCISE 16.4

1. Evaluate the following: (a) $2 \log 1\frac{1}{3} + \log 1\frac{1}{2} - 3 \log 2 + \log 3$ (b) $\log 2\frac{2}{5} + \log 2\frac{1}{7} - \log 5\frac{1}{7}$

2. Solve the equation $\log_{10} x = 1 + \frac{3}{2} \log_{10} 4$

3. Solve the equations: (a) $4^x = 16^{x+1}$, (b) $3^x = 8^{x+1}$

4. Solve the equation $2^{x^2+8} = 4^{3x}$

5. Solve the equation $25^x - 5^{x+2} + 100 = 0$

6. Solve the simultaneous equations $\log_3(3x + y) = 2$
$$\log_2(2y - x) = 2$$

SUMMARY

1. The laws of indices state that for all m and n

(a) $a^m \times a^n = a^{m+n}$, (b) $a^m \div a^n = a^{m-n}$,

(c) $(a^m)^n = a^{mn}$

Also $a^{\frac{1}{n}} = \sqrt[n]{a}, \sqrt[n]{a^m} = a^{\frac{m}{n}}, a^{-n} = \frac{1}{a^n}, a^0 = 1$

2. The laws of logarithms state that (a) $\log x + \log y = \log xy$

(b) $\log x - \log y = \log \frac{x}{y}$, (c) $n \log x = \log x^n$

Also $\log_a a = 1, \log_a 1 = 0, \log \frac{1}{x} = -\log x$

3. Logarithms can be used to solve equations when the unknown is an index.

SELF ASSESSMENT PAPER No 16

Instructions: Answer all questions in both sections
Time allowed: Section A 30 minutes
　　　　　　　　(10 marks each question)
　　　　　　　　Section B 30 minutes
　　　　　　　　(20 marks each question)
Marks gained: 45+ pass with credit, 36–45 pass, less than 36 fail, repeat chapter 16

Section A

1. Find the values of (a) $8^{\frac{1}{3}}$, (b) $81^{\frac{1}{4}}$, (c) $(7\frac{1}{9})^{\frac{1}{2}}$

2. Reduce to a single logarithm $\log 3 - 2 \log 2 + \log \frac{8}{3}$

3. Find the values of (a) $\log_2 8$, (b) $\log_3 9$

4. Solve the equation $3^x = 9^{2x-5}$

5. Solve the equation $2^x = 5$

Section B

1. (a) Evaluate $\log 1\frac{1}{2} - \log 3\frac{1}{2} + \log 2\frac{1}{3}$

　(b) Solve the equation $2^{x-1}.3^x = 5^{x+2}$

2. (a) Solve the equation $\log_3(x + 2) = 2$

　(b) Solve the equation $4^{(3x^2+8)} = 16^{5x}$

ANSWERS

Exercise 16.1　　1. (a) 8, (b) 9, (c) 2　　　2.　(a) 0.04, (b) 0.111, (c) 0.031 25, (d) 0.5,　　3. (a) 3.66, (b) 0.668, (c) 5.20,　　4.　1, 5.12, 6.2

Exercise 16.2　1.　$\log \frac{3}{4}$,　　2.　$\log 9$,　3. $\log 6$, 4.　$\log \frac{3}{8}$,　　5. (a) 3, (b) 1,　　6. (a) 3, (b) 5, 7.　$1\frac{1}{2}$,　　8. (a) 2, (b) 3

Exercise 16.3　1. (a) 2.578, (b) 1.043, (c) 3.174, (d) 3.743,　　2. (a) 2.391, (b) 2.135, 3.　$- 1.865$,　　4. 0.24

Exercise 16.4　　1. (a) 0, (b) 0　　2. 80,　3. (a)　2, (b) $- 2.12$,　　4. 2, 4,　　5. 1.861, 1, 6. $x = 2, y = 3$

SELF ASSESSMENT PAPER No 16

Marks

Section A

1. (a) 2   3
   (b) 3   3
   (c) $\left(\dfrac{64}{9}\right)^{\frac{1}{2}}$ 2

   $= 2\frac{2}{3}$ 2

2. $\log \dfrac{3}{2^2} \times \dfrac{8}{3}$ 6

   $= \log 2$ 4

3. (a) $3 \log_2 2$ 3
   $= 3$ 2
   (b) $2 \log_3 3$ 3
   $= 2$ 2

4. $3^x = 3^{2(2x-5)}$ 3
   $x = 2(2x-5)$ 3
   $3x = 10$ 2
   $x = 3\frac{1}{3}$ 2

5. $x \log_{10} 2 = \log_{10} 5$ 4

   $x = \dfrac{\log_{10} 5}{\log_{10} 2}$ 4

   $x = 2.322$ 2

Section B

1. (a) $\log \dfrac{3}{2} \times \dfrac{1}{\frac{7}{2}} \times \dfrac{7}{3}$ 4

   $= \log 1$ 3
   $= 0$ 1
   (b) $\log_{10} 2^{x-1}.3^x = \log_{10} 5^{x+2}$ 2
   $(x-1)\log_{10} 2 + x\log_{10} 3 = (x+2)\log_{10} 5$ 4
   $0.3010(x-1) + 0.4771x = 0.699(x+2)$ 3
   $0.0791x = 1.699$ 2
   $x = 21.5$ 1

2. (a) $x + 2 = 3^2$ 4
   $x + 2 = 9$ 1
   $x = 7$ 1
   (b) $4^{(3x^2+8)} = 4^{2 \times 5x}$ 4
   $3x^2 + 8 = 10x$ 2
   $3x^2 - 10x + 8 = 0$ 2
   $(3x-4)(x-2) = 0$ 4
   $x = 1\frac{1}{3}$ or $x = 2$ 2

PHASE TEST 5

Instructions: Answer all questions
Time allowed: Section A 25 minutes (30 marks)
               Section B 35 minutes
               (15 marks each question)
Marks gained: 37+ pass with credit, 30–37 pass, less than 30 fail, repeat chapters 15 and 16

Section A

1. Find the factors of (i) $3x^2 - 7x + 4$, (ii) $2x^2 - 18$
2. Solve the equations (i) $3x^2 - 11x - 4 = 0$
                     (ii) $4x^2 + 7x + 2 = 0$
3. Write as a number (i) $\log_2 4$, (ii) $\log_3 27$
4. Write as a single logarithm $2 \log 7 - \log 63 + 3 \log 3$

Section B

1. The power P watts developed in an electrical circuit is given by $P = 9I - 7I^2$, where I is the current in amperes. Determine the current when a power of 2.3 W is produced in the circuit.

2. Solve the simultaneous equations $x + y = 4$
   $$2x^2 - 5x + y = 4$$

3. Solve the equations $3^{x+1}.2^x = 5^{x-2}$

ANSWERS PHASE TEST 5

Marks

Section A

1. (i) $(3x-4)(x-1)$ 2
   (ii) $2(x^2 - 9)$ 1
         $2(x-3)(x+3)$ 2

2. (i) $(3x+1)(x-4) = 0$ 2
        $3x + 1 = 0$ or $x - 4 = 0$ 1
        $x = -\frac{1}{3}$ or $x = 4$ 1
   (ii) $x = \dfrac{-7 \pm \sqrt{7^2 - 4 \times 4 \times 2}}{2 \times 4}$ 4

   $x = \dfrac{-7 \pm \sqrt{17}}{8} = \dfrac{-7 \pm 4.123}{8}$ 2

   $x = -1.39$ or $x = -0.36$ 2

3. (i) $\log_2 2^2 = 2 \log_2 2$ 2
             $= 2$ $(\log_2 2 = 1)$ 2
   (ii) $\log_3 3^3 = 3 \log_3 3$ 2
             $= 3$ $(\log_3 3 = 1)$ 2

4. $\log 7^2 - \log 63 + \log 3^3$ 2
   $\log \dfrac{7 \times 7 \times 3 \times 3 \times 3}{63}$ 2

   $\log 21$ 1

Section B

1. $9I - 7I^2 = 2.3$ 4
   $7I^2 - 9I + 2.3 = 0$ 1
   $I = \dfrac{9 \pm \sqrt{81 - 4 \times 7 \times 2.3}}{2 \times 7}$ 4

   $I = \dfrac{9 \pm 4.07}{14}$ 3

   $I = 0.934$ A, or 0.352 A   3

	Marks
2. $y = 4 - x$	1
$2x^2 - 5x + 4 - x = 4$	4
$2x^2 - 6x = 0$	1
$2x(x - 3) = 0$	3
$x = 0$ or $x = 3$	2
When $x = 0, y = 4$	2
When $x = 3, y = 1$	2

	Marks
3. $\log_{10} 3^{x+1} + \log_{10} 2^x = \log_{10} 5^{x-2}$	4
$(x + 1) \log_{10} 3 + x \log_{10} 2 = (x - 2) \log_{10} 5$	3
$0.4771(x + 1) + 0.3010x = 0.699(x - 2)$	3
$0.079x = -1.8751$	4
$x = -23.7$	1

17 Conic Sections and Non Linear Laws

17.1 CONIC SECTIONS

Consider a hollow cone, as shown in fig. 17.1, extended to infinity in both directions. AOB and COD are called generating lines of the cone. If ZOZ is the axis of the cone then:

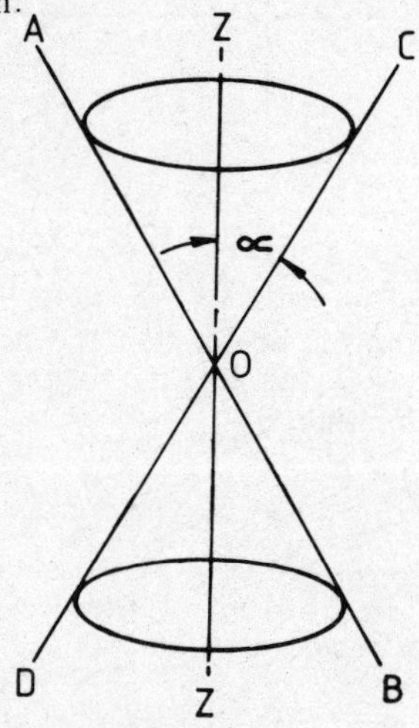

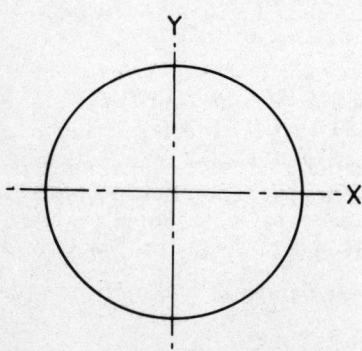

Figure 17.1

(1) A plane at right angles to ZOZ cuts the cone in a circle. The circle is usually drawn with its centre as the origin of co-ordinates, see fig. 17.2. Its equation is then $x^2 + y^2 = r^2$, where r is the radius of the circle.

Figure 17.2

(2) A plane making an angle greater than α with ZOZ cuts the cone in an ellipse. This is usually drawn with its centre at the origin of co-ordinates, see fig. 17.3. The equation of the ellipse is then $\dfrac{x^2}{a^2} + \dfrac{y^2}{b^2} = 1$

where a and b are the semi-major and semi-minor axes.

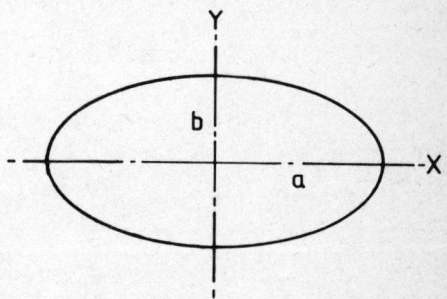

Figure 17.3

(3) A plane parallel to a generating line cuts the cone in a parabola. This can be drawn as shown in fig. 17.4, with the vertex at the origin of co-ordinates. The equation is then $y^2 = 4ax$, where a is a constant. We will look at other parabolas, like $y = ax^2 + bx + c$ later.

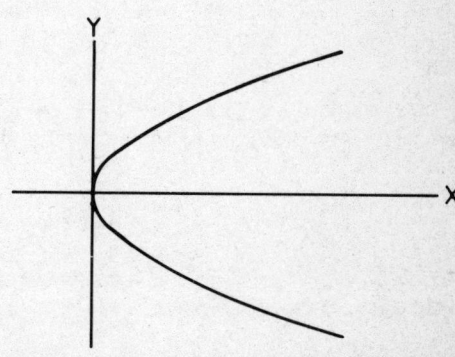

Figure 17.4

(4) Any other plane cuts the cone in a hyperbola. This is usually drawn as in fig. 17.5 when its equation is in the form $\dfrac{x^2}{a^2} - \dfrac{y^2}{b^2} = 1$ where a and b are constants.

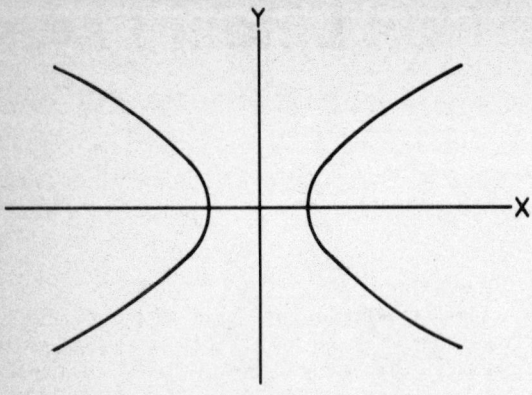

Figure 17.5

Sometimes the hyperbola is drawn as in fig. 17.6, this occurs when the temperature of a gas is kept constant the pressure p, and volume v, are related by $pv =$ constant. This curve is called a rectangular hyperbola.

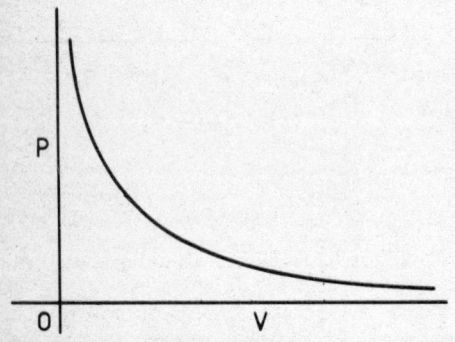

Figure 17.6

We have already looked at the curve for a quadratic function this is a parabola, $y = ax^2 + bx + c$, see chapter 3.

Example 17.1 Draw the graph of $y = x^2 + 6x + 2$. On the same axes draw the graphs of $y = -x^2 + 6x + 2$ and $y = x^2 + 4x + 2$.

x	-3	-2	-1	0	1	2	3	4
$y = x^2 + 2x + 6$	9	6	5	6	9	14	21	30
$y = -x^2 + 2x + 6$	-9	-2	3	6	7	6	3	-2
$y = x^2 + 4x + 6$	3	2	3	6	11	18	27	38

From the graphs, in fig. 17.7, it can be seen that the parabola $y = x^2 + 2x + 6$ has a minimum value of 5 at $x = -1$.
By changing the sign of x^2 the parabola is turned upside down and now has a maximum of 7 at the point $x = 1$.
This is true for all parabolas, $y = ax^2 + bx + c$ when the a is negative the parabola has a maximum value,

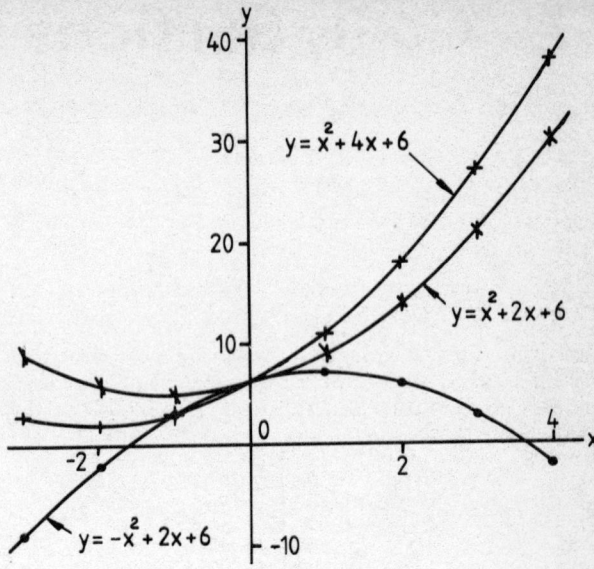

Figure 17.7

when the a is positive the parabola has a minimum value. Increasing the size of the term in x, increases the slope of the curve and also moves the position of the maximum or minimum.

EXERCISE 17.1

1. Draw the graph of $y = 2x^2 - 3x + 3$ and $y = 2x^2 + 3x + 3$, for values of x from -4 to $+4$. Compare the two parabolas when the x term changes sign.

2. Draw the graph of $y = \dfrac{2}{x}$ for values of x from -4 to $+4$.
Note how the value of y changes between -1 and $+1$. The value at $x = 0$ will be infinite. When x is very small and positive y is very large and positive. When x is very small and negative, y is very large and negative.

3. Draw the graph of $y = 3\sqrt{x}$ for values of x from 0 to 6. Note there are no values of y when x is negative.

4. Draw the graph of $y = 2x^2 - 5x + 2$ and $y = 2x^2 - 5x + 6$, for values of x from -4 to $+4$. Compare the two parabolas when the constant changes from 2 to 6.

17.2 NON LINEAR LAWS REDUCIBLE TO A STRAIGHT LINE

Certain relations, if plotted as graphs, would not produce a straight line. For example $y = ax^2 + b$ would produce a parabola. But if we plot y against x^2 we will obtain a straight line. Similarly we can plot $y = \dfrac{a}{x} + b$ on a graph as y against $\dfrac{1}{x}$ and we will again obtain a straight line.

Example 17.2 The following results were obtained in measuring the inductance L of a coil of t turns.

t	5	6	7	8	9	10
L	21	27	36	44	55	67

By plotting L against t^2 show that these results satisfy a law of the form $L = at^2 + b$. From your graph, find suitable values of a and b.

L	21	27	36	44	55	67
t^2	25	36	49	64	81	100

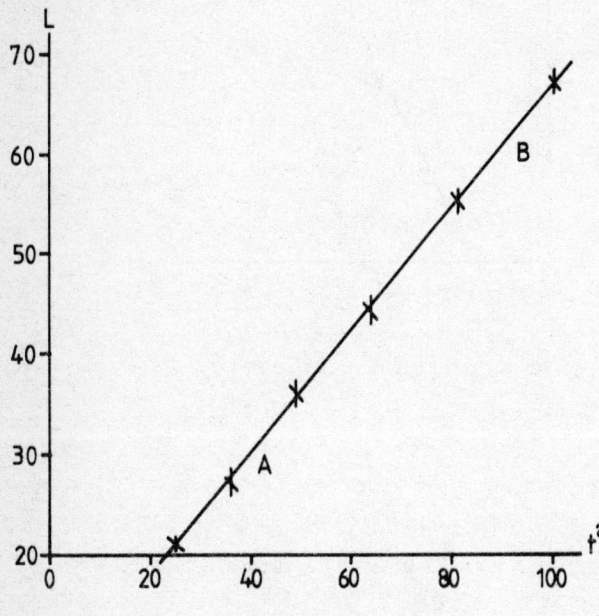

Figure 17.8

The graph is drawn in fig. 17.8, not all the points lie exactly on the line drawn. The best straight line is drawn through the experimental points. It is recommended that you draw your own graph and read off points for the solution, you can then compare your answer with the working below.

Taking two points from the graph, A and B

At A, $L = 30$, $t^2 = 40$; $30 = 40a + b$ (1)

At B, $L = 60.5$, $t^2 = 90$; $60.5 = 90a + b$ (2)

To solve these equations subtract (1) from (2)

$$30.5 = 50a, \quad a = \frac{30.5}{50} = 0.61$$

Substituting this value of a into (1) $30 = 40 \times 0.61 + b$

$$b = 30 - 40 \times 0.61 = 30 - 24.4 = 5.6$$

The results satisfy the law $L = 0.61t^2 + 5.6$ because all of the points lie approximately on this line.

Other laws of this type are:

$y = ax^3 + b$, plot y against x^3

$y = \dfrac{a}{x} + b$, plot y against $\dfrac{1}{x}$

$y = ax^2 + bx$, plot $\dfrac{y}{x}$ against x

(This law can be written $\dfrac{y}{x} = ax + b$ by dividing through by x.)

EXERCISE 17.2

1. In an experiment the following values of L and M were obtained:

L	0.200	0.091	0.067	0.050	0.040
M	2.1	3.0	3.6	4.3	5.0

These values are believed to be connected by the equation $M = a + \dfrac{b}{L}$. Draw a graph to test the assumption and find the values of a and b.

2. Verify that the values of two observed quantities x and y satisfy a law of the form $y = a + bx^2$ and find suitable values for a and b.

x	1.1	1.8	2.5	2.9	3.6	4.3	4.8
y	1.91	2.13	2.42	2.65	3.09	3.66	4.09

3. The law connecting the horse-power H required to drive a certain vessel at a speed of V knots is $H = aV^3 + b$, where a and b are constants. Actual values of H and V obtained during a trial run are:

H	290	560	1144	1810	2300
V	5	7	9	11	12

Test whether these values are in accordance with the stated law, and, if so, determine the values of the constants a and b.

4. The following table gives experimental values of two variables x and y assumed to be connected by the law

$$\frac{y}{x} = a + bx$$

x	2	4	6	8	10
y	11.2	12.7	4.9	− 13.0	− 40.0

Verify the law and determine values for a and b.

5. The current I mA through a rectifier is measured for values of the applied voltage V volts.

V (V)	8	12	18	25	32
I mA)	3.4	5.9	10.5	17.9	26.8

Assuming the law $I = aV + bV^2$ applies to these plot suitable variables to obtain a straight line graph. Hence find the values for the constants a and b.

7. Readings of oil consumption p at a bearing of an engine at various temperatures T of the bearing are as follows:

T	0	10	15	18	24	30
p	0.5	0.592	0.767	0.913	1.28	1.72

Test whether these observations, within reasonable experimental limits, are consistent with the relationship $p^2 = aT^3 + b$, and if found to be so, determine the values of the constants a and b.

17.3 LAWS REQUIRING LOGARITHMS

In laws of the type $y = ax^n$ logarithms must be taken of both sides to obtain an equation which can be compared with a straight line equation.

$$y = ax^n$$

Taking logarithms of both sides of this equation

$$\log y = \log ax^n = \log a + \log x^n$$

$$\log y = \log a + n \log x$$

$$\text{or } \log y = n \log x + \log a$$

The variables are now $\log y$ and $\log x$, and if we plot $\log y$ against $\log x$ we will obtain a straight line, the gradient will be n and the intercept on the y-axis will be $\log a$. It is usual to use logarithms to base 10, since these are available in tables and on most calculators.

Example 17.3 The following measurements of the luminosity of an electric lamp with voltage (V) were obtained.

V	60	70	80	90	100	110
L	6.5	12.1	20.5	33.3	50	75.9

The law is thought to be of the form $L = aV^n$. Verify this and determine values for a and n.

$$L = aV^n$$

By taking logarithms of both sides as before we have

$$\log L = n \log V + \log a$$

We therefore plot $\log_{10} L$ against $\log_{10} V$

$\log_{10} L$	0.813	1.083	1.312	1.522	1.699	1.880
$\log_{10} V$	1.778	1.845	1.903	1.954	2.000	2.041

Since a straight line can be drawn through the points plotted on the graph in fig. 17.9 the law is verified. Taking two points from the graph

When $\log_{10} V = 1.8$, $\log_{10} L = 0.9$; $0.9 = 1.8n + \log_{10}a$
$$(1)$$
When $\log_{10} V = 2.0$, $\log_{10} L = 1.72$; $1.72 = 2.0n + \log_{10}a$
$$(2)$$

The values must be taken from the graph not from the table of results since it is the law of the line we have drawn that we require.

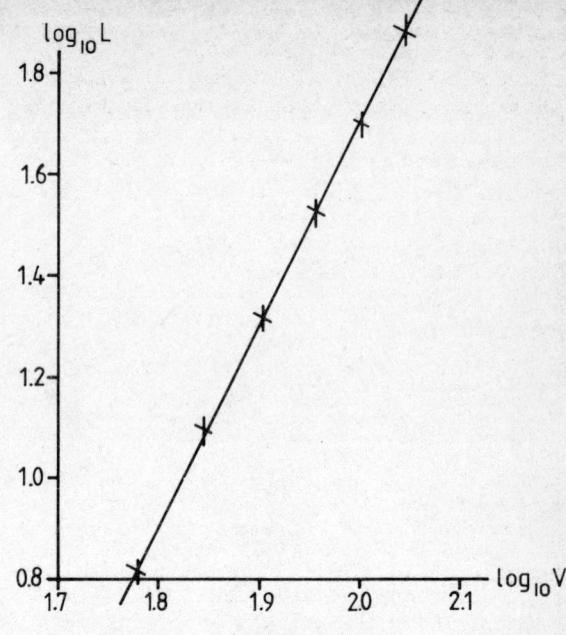

Figure 17.9

Take equation (1) from equation (2)

$$0.82 = 0.2n, \; n = \frac{0.82}{0.2} = 4.1$$

Substituting $n = 4.1$ into (1) $0.9 = 1.8 \times 4.1 + \log_{10}a$

$$\log_{10}a = 0.9 - 1.8 \times 4.1 = -6.48$$

To find a we must find the antilogarithm of -6.48
It is easier to work out if we write this as

$$-6.48 = -7 + 0.52$$

The value of a is then the antilogarithm of 0.52×10^{-7}
The antilogarithm of 0.52 is 3.31

$$a = 3.31 \times 10^{-7}$$

The law is $L = (3.31 \times 10^{-7}) V^{4.1}$

A very similar law occurs when a gas is compressed. The law is $pV^n = c$
When we take logarithms of both sides we have
$\log pV^n = \log c$

$$\log p + \log V^n = \log c$$

$$\log p = -n \log V + \log c$$

We plot $\log p$ against $\log V$, the straight line will usually have a negative slope.

EXERCISE 17.3

1. The following table gives corresponding values of x and y.

x	3	5	7	9	11	13
y	18.2	39.1	64.8	94.5	127.7	164.1

Plot a suitable graph to verify that these values are connected by the law $y = ax^n$. What are the values of the constants a and n?

2. The table below shows the pressure and volume of a quantity of air when subjected to a moderately rapid compression.

p	15	30	60	100	200	400
V	10.00	5.67	3.15	2.06	1.15	0.65

Show that p and V are connected by the law $pV^n = k$ and find the constants n and k.

3. The following values of current I and voltage V were observed in a certain circuit.

I	1.0	2.6	8.8	18.2	30.6
V	0.6	1.0	2.0	3.0	4.0

Draw a graph to verify that these follow a law of the form $I = aV^n$ and use it to find values for a and n.

4. A gas is compressed in a cylinder by the movement of a piston. The relation between the force F in the piston and the distance D from the inside face of the piston to the inside end of the cylinder is thought to be of the type $FD^n = $ a constant. The following values were obtained experimentally:

F	110	142	198	294	523	1406
D	18	15	12	9	6	3

Verify the law, and determine the best values of the constants.

17.4 GENERAL EXAMPLES

Example 17.4 Verify graphically that the following corresponding values of two variables V and h are related by the law $V = h(ah^2 + b)$ where a and b are constants determine from the graph the values of a and b.

V	0.20	0.45	0.88	1.55	2.52
h	1	1.5	2	2.5	3

Find the value of V when $h = 1.75$.
Dividing both sides of $V = h(ah^2 + b)$ by h we have

$$\frac{V}{h} = ah^2 + b$$

To obtain a straight line graph we must plot $\frac{V}{h}$ against h^2.

$\frac{V}{h}$	0.20	0.30	0.44	0.62	0.84
h^2	1	2.25	4.00	6.25	9.00

The graph is drawn in fig. 17.10. In this case the value of b can be read directly from the graph, b is the intercept on the $\frac{V}{h}$ axis

$$b = 0.12$$

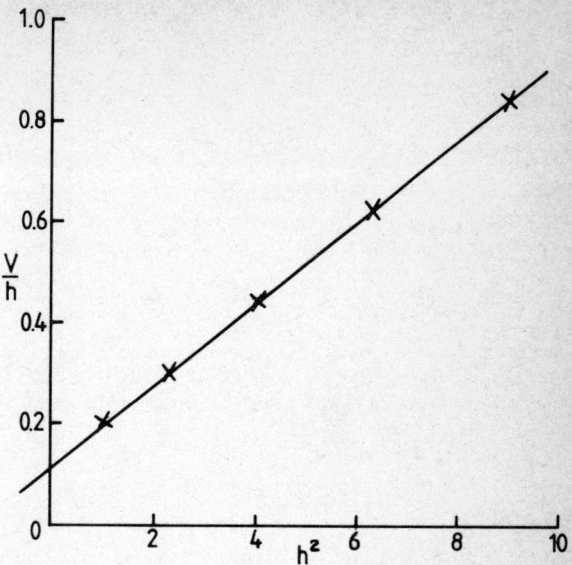

Figure 17.10

The gradient is $a = \dfrac{0.4}{5.0} = 0.08$

The law is $V = h(0.08h^2 + 0.12)$

The value of V when $h = 1.75$ can be found from the graph or from the law. From the graph when $h = 1.75$, $h^2 = 3.06$,

$$\frac{V}{h} = 0.365, \quad V = 0.365 \times 1.75 = 0.639$$

From the law $V = 1.75(0.08 \times 1.75^2 + 0.12)$

$$V = 0.639$$

When $h = 1.75$, $V = 0.639$

EXERCISE 17.4

1. On the same axes and with the same scales plot the graphs of $y = 3 + 2x - x^2$ and $y = \dfrac{2}{x}$ from $x = -3$ to $x = +3$. From the graphs obtain the three roots of the equation

$$x^3 - 2x^2 - 3x + 2 = 0$$

2. Two quantities W and H are believed to be connected by a law of the form: $W = \dfrac{m}{\sqrt{H}} + c$.
The following values of W and H were obtained by experiment.

W	5.02	4.50	4.13	4.04	3.97	3.92
H	4	6	8	10	12	14

Draw a suitable graph to test if W and H are connected by the law, and determine the probable values of m and c. Using the law, find the value of H when $W = 4.25$.

3. Verify that the following values of x and y are connected by the equation $y = \dfrac{ax}{x + b}$ and determine the best values of a and b.

x	0	1	2	3	4	5
y	0	0.39	0.67	0.86	1.02	1.11

4. The following table gives the safe load L carried by steel troughing of a particular section for different spans S

S	10	14	18	22	26
L	44.6	22.5	13.6	9.1	6.5

Draw a graph to test if the law connecting L and S is of the form $L = kS^n$ where k and n are constants. Use the graph to find values of k and n.
Find the value of L when $S = 20$.

SUMMARY

1. When a cone is cut by a plane the section is called a conic section. Depending on the angle of the plane the section is a circle, ellipse, parabola or an hyperbola. These are sketched in figs. 17.2 to 17.6.

2. The equation $y = ax^2 + bx + c$ is a parabola. When a is positive the parabola has a minimum value, when a is negative the parabola has a maximum value.

3. The graphs of equations such as (1) $y = ax^2 + b$, (2) $y = a\sqrt{x} + b$, are curves but if in (1) y is plotted against x^2 and in (2) y is plotted against $\sqrt{x}$ then the graph becomes a straight line. a is the gradient of the line and b is the intercept on the y-axis.

4. To determine the law $y = ax^n$, logarithms are taken of both sides $\log y = n \log x + \log a$. $\log y$ is plotted against $\log x$. The gradient is n and the intercept on the y-axis is $\log a$.

5. To determine the law $pV^n = c$, logarithms of both sides are taken $\log p = -n \log V + \log c$. $\log p$ is plotted against $\log V$. The gradient is usually negative. $-n$ is the gradient and $\log c$ is the intercept on the y-axis.

SELF ASSESSMENT PAPER No 17

Instructions: Answer all questions
Time allowed: One and a half hours
(20 marks each question)
Marks gained: 30+ pass with credit, 24–30 pass, less than 24 fail, repeat chapter 17

1. (a) Sketch the graph $\dfrac{x^2}{9} + \dfrac{y^2}{4} = 1$ for values of x from -3 to $+3$.

(b) Draw the graph of $y = x^2 - x + 6$ for values of x from -4 to $+4$. Write down the equation of a similar parabola but which has a maximum value.

2. The law connecting R and V is believed to be of the form $R = aV^2 + b$. With the values given below test if this is true and determine the best values of a and b.

V	12	16	20	22	24	26
R	6.45	7.55	9.02	9.85	10.77	11.74

3. Air is being compressed and the pressure and volume are recorded. The law is thought to be of the form $pV^n = c$. Verify this law from the following results and evaluate n and c.

Pressure p	1.92	2.80	5.00	8.30	25.00
Volume V	26	18	10	6	2

ANSWERS

Exercise 17.2 1. $a = 1.4, b = 0.145$, 2. $a = 1.8, b = 0.10$, 3. $a = 1.26, b = 132$, 4. $a = 8, b = -1.2$, 5. $a = 0.278, b = 0.0174$, 6. $a = 0.0001, b = 0.25$

Exercise 17.3 1. $n = 1.5, a = 3.5, k = 238$, 2. $n = 1.2$, 3. $a = 2.45, n = 1.88$, 4. $n = 1.42$, constant $= 6600$

Exercise 17.4 1. $-1.4, 0.6, 2.8$, 2. $m = 4.9, c = 2.53, H = 8.1$, 3. $a = 2.21, b = 4.66$, 4. $n = -2, k = 4440, L = 11.1$

SELF ASSESSMENT PAPER No 17

Marks

1. (a) Graph, ellipse cutting x-axis at -3 and $+3$ and y-axis at -2 and $+2$ — 8
(b) Graph, parabola cutting y-axis at 6 — 8
$y = -x^2 - x - 6$ has a maximum value — 4

2. Plot R against V^2 — 2

R	6.45	7.55	9.02	9.85	10.77	11.74
V^2	144	256	400	484	576	676

— 6

Graph of R against V^2 — 6
$a = 0.01$ — 3
$b = 5.01$ — 3

3. $\log_{10} p = -n \log_{10} V + \log_{10} c$ — 2
Plot $\log_{10} p$ against $\log_{10} V$ — 1

$\log_{10} p$	0.2833	0.4472	0.6990	0.9191	1.3979
$\log_{10} V$	1.4150	1.2553	1.0000	0.7782	0.3010

— 3
— 3

Graph of $\log_{10} p$ against $\log_{10} V$ — 4
Graph is a straight line, law is verified — 1
$n = 1.0$ — 3
$\log_{10} c = 1.7$ — 2
$c = 50.1$ — 1

18 Matrices and Determinants

18.1 MATRIX ADDITION

Suppose a certain industrial organisation has three factories all making components A, B, C and D. The stock at these factories could be summarised as follows:

	A	B	C	D
Factory 1	10	5	9	7
Factory 2	3	12	7	8
Factory 3	5	9	7	6

This display is called a matrix, the matrix being

$$\begin{pmatrix} 10 & 5 & 9 & 7 \\ 3 & 12 & 7 & 8 \\ 5 & 9 & 7 & 6 \end{pmatrix}$$

This matrix is said to be a 3 by 4 matrix since it has 3 rows and 4 columns. In general a matrix is an m by n matrix having m rows and n columns.

The number of components used could be represented by the matrix

$$\begin{pmatrix} 2 & 1 & 3 & 1 \\ 3 & 5 & 2 & 2 \\ 1 & 3 & 2 & 1 \end{pmatrix}$$

and the number of components still remaining in stock could be found by subtracting corresponding entries in the two matrices

$$\begin{pmatrix} 10-2 & 5-1 & 9-3 & 7-1 \\ 3-3 & 12-5 & 7-2 & 8-2 \\ 5-1 & 9-3 & 7-2 & 6-1 \end{pmatrix} = \begin{pmatrix} 8 & 4 & 6 & 6 \\ 0 & 7 & 5 & 6 \\ 4 & 6 & 5 & 5 \end{pmatrix}$$

Hence we define addition or subtraction of matrices as the addition or subtraction of corresponding entries. To be able to do this the matrices must have the same number of rows and columns. Even though there were no components A left in factory 2 we must put in a zero.

If all of the elements are zero, the matrix is called a zero matrix.

Example 18.1 $A = \begin{pmatrix} 3 & 5 \\ -2 & 4 \end{pmatrix}$, $B = \begin{pmatrix} 0 & 4 \\ 3 & -1 \end{pmatrix}$

Find (a) $A + B$, (b) $A - B$, (c) $2A$, (d) $4B$

(a) $A + B = \begin{pmatrix} 3+0 & 5+4 \\ -2+3 & 4-1 \end{pmatrix} = \begin{pmatrix} 3 & 9 \\ 1 & 3 \end{pmatrix}$

(b) $A - B = \begin{pmatrix} 3-0 & 5-4 \\ -2-3 & 4-(-1) \end{pmatrix} = \begin{pmatrix} 3 & 1 \\ -5 & 5 \end{pmatrix}$

(c) $2A = A + A = \begin{pmatrix} 6 & 10 \\ -4 & 8 \end{pmatrix}$

Here we see that $2A$ is found by multiplying each element by 2.

(d) $4B = 4 \times \begin{pmatrix} 0 & 4 \\ 3 & -1 \end{pmatrix} = \begin{pmatrix} 0 & 16 \\ 12 & -4 \end{pmatrix}$

EXERCISE 18.1

1. Given $A = \begin{pmatrix} 3 & 1 \\ 4 & 2 \end{pmatrix}$, $B = \begin{pmatrix} 6 \\ 7 \end{pmatrix}$, $D = \begin{pmatrix} 3 & 2 \\ 1 & 5 \end{pmatrix}$

$E = (2 \ \ 1)$, $F = (3 \ \ 7)$, $G = \begin{pmatrix} 0 \\ 0 \end{pmatrix}$

Find the sum of each pair of matrices that can be added together.

2. When $A = \begin{pmatrix} 3 & 2 \\ 5 & 1 \end{pmatrix}$, $B = \begin{pmatrix} -1 & 3 \\ -4 & 2 \end{pmatrix}$, $C = \begin{pmatrix} 2 & -1 \\ -3 & 4 \end{pmatrix}$

find (a) $A + B$, (b) $A - B$, (c) $B + C$, (d) $A - C$, (e) $2A$, (f) $3C$, (g) $2A - 3B$, (h) $2C - 4B$

3. Three garages sell cars, vans and motorcycles. At the beginning of one week they have in stock the following:
garage 1 has 7 cars, 3 vans and 2 motorcycles; garage 2 has 8 cars, 5 vans and 4 motorcycles; garage 3 has 11 cars, 3 vans and 5 motorcycles. Represent these by a 3 by 3 matrix. During the week garage 1 sells 3 cars, 1 van and 1 motorcycle; garage 2 sells 3 cars, 3 vans and 2 motorcycles and garage 3 sells 5 cars and 2 motorcycles.
Represent the sales by a matrix and also the stocks at the end of the week.

18.2 MATRIX MULTIPLICATION

We have already seen that if we multiply a matrix by a constant then we must multiply each element in the matrix by that constant.

Thus if $A = \begin{pmatrix} 1 & 6 \\ 2 & 0 \end{pmatrix}$, $3A = \begin{pmatrix} 3 & 18 \\ 6 & 0 \end{pmatrix}$

If in paragraph 18.1 the component A has a mass 2 kg, the component B 3 kg, the component C 1 kg and the component D 4 kg then the toal mass of the components in factory 1 is $(10 \times 2 + 5 \times 3 + 9 \times 1 + 7 \times 4)$ kg or 72 kg etc.

Hence we could write this

$$\begin{pmatrix} 10 & 5 & 9 & 7 \\ 3 & 12 & 7 & 8 \\ 5 & 9 & 7 & 6 \end{pmatrix} \begin{pmatrix} 2 \\ 3 \\ 1 \\ 4 \end{pmatrix} = \begin{pmatrix} 10 \times 2 + 5 \times 3 + 9 \times 1 + 7 \times 4 \\ 3 \times 2 + 12 \times 3 + 7 \times 1 + 8 \times 4 \\ 5 \times 2 + 9 \times 3 + 7 \times 1 + 6 \times 4 \end{pmatrix}$$

$$= \begin{pmatrix} 72 \\ 81 \\ 68 \end{pmatrix}$$

If component A cost £50, component B costs £30, component C costs £70 and component D costs £20, the total cost of the components in factory 1 would be £$(10 \times 50 + 5 \times 30 + 9 \times 70 + 7 \times 20) = $ £1420
The cost could be joined with the mass to form a 4 by 2 matrix

$$\begin{pmatrix} 10 & 5 & 9 & 7 \\ 3 & 12 & 7 & 8 \\ 5 & 9 & 7 & 6 \end{pmatrix} \begin{pmatrix} 2 & 50 \\ 3 & 30 \\ 1 & 70 \\ 4 & 20 \end{pmatrix} = \begin{pmatrix} 72 & 1420 \\ 81 & 1160 \\ 68 & 1130 \end{pmatrix}$$

We thus see that in matrix multiplication every element of the first row of the first matrix is multiplied by the corresponding element in the first column of the second matrix, the sum forms the first element of the product. Then each element of the first row of the first matrix is multiplied by the corresponding element in the second column of the second matrix and the sum forms the second element in the first row of the product etc.

We therefore see that for multiplication the first matrix must have the same number of columns as there are rows in the second matrix.

Example 18.2 Find the matrix products where possible of the following:

$A = \begin{pmatrix} 2 & 1 \\ -1 & 3 \end{pmatrix}$, $B = \begin{pmatrix} 3 & -2 \\ 4 & 0 \end{pmatrix}$, $C = \begin{pmatrix} 2 \\ 4 \end{pmatrix}$

$D = \begin{pmatrix} 5 & 7 & 2 \\ 0 & 1 & 3 \end{pmatrix}$, $E = (1 \quad 3)$

$A.B = \begin{pmatrix} 2 & 1 \\ -1 & 3 \end{pmatrix}\begin{pmatrix} 3 & -2 \\ 4 & 0 \end{pmatrix} = \begin{pmatrix} 2 \times 3 + 1 \times 4 & 2 \times -2 + 1 \times 0 \\ -1 \times 3 + 3 \times 4 & -1 \times -2 + 3 \times 0 \end{pmatrix}$

$$= \begin{pmatrix} 10 & -4 \\ 9 & 2 \end{pmatrix}$$

$B.A = \begin{pmatrix} 3 & -2 \\ 4 & 0 \end{pmatrix}\begin{pmatrix} 2 & 1 \\ -1 & 3 \end{pmatrix} = \begin{pmatrix} 8 & -3 \\ 8 & 4 \end{pmatrix}$

We notice that $A.B \neq B.A$ hence matrix multiplication is not commutative.

$A.C = \begin{pmatrix} 2 & 1 \\ -1 & 3 \end{pmatrix}\begin{pmatrix} 2 \\ 4 \end{pmatrix} = \begin{pmatrix} 8 \\ 10 \end{pmatrix}$　$C.A$　does not exist

$A.D = \begin{pmatrix} 2 & 1 \\ -1 & 3 \end{pmatrix}\begin{pmatrix} 5 & 7 & 2 \\ 0 & 1 & 3 \end{pmatrix} = \begin{pmatrix} 10 & 15 & 7 \\ -5 & -4 & 7 \end{pmatrix}$

$D.A$ does not exist

$A.E$ does not exist

$E.A = (1 \quad 3) \begin{pmatrix} 2 & 1 \\ -1 & 3 \end{pmatrix} = (-1 \quad 10)$

$B.C = \begin{pmatrix} 3 & -2 \\ 4 & 0 \end{pmatrix}\begin{pmatrix} 2 \\ 4 \end{pmatrix} = \begin{pmatrix} -2 \\ 8 \end{pmatrix}$, $C.B$ does not exist

$B.D = \begin{pmatrix} 3 & -2 \\ 4 & 0 \end{pmatrix}\begin{pmatrix} 5 & 7 & 2 \\ 0 & 1 & 3 \end{pmatrix} = \begin{pmatrix} 15 & 19 & 0 \\ 20 & 28 & 8 \end{pmatrix}$

$D.B$ does not exist

$B.E$ does not exist

$E.B = (1 \quad 3)\begin{pmatrix} 3 & -2 \\ 4 & 0 \end{pmatrix} = (15 \quad -2)$

$C.D$ and $D.C$ do not exist

$C.E = \begin{pmatrix} 2 \\ 4 \end{pmatrix}(1 \quad 3) = \begin{pmatrix} 2 & 6 \\ 4 & 12 \end{pmatrix}$, $E.C$ does not exist

$D.E$ does not exist,

$E.D = (1 \quad 3)\begin{pmatrix} 5 & 7 & 2 \\ 0 & 1 & 3 \end{pmatrix} = (5 \quad 10 \quad 11)$

Example 18.3 When $A = \begin{pmatrix} 1 & 0 \\ 0 & 1 \end{pmatrix}$, $B = \begin{pmatrix} 3 & 2 \\ 1 & -3 \end{pmatrix}$

find $A.B$ and $B.A$

$A.B = \begin{pmatrix} 1 & 0 \\ 0 & 1 \end{pmatrix}\begin{pmatrix} 3 & 2 \\ 1 & -3 \end{pmatrix} = \begin{pmatrix} 3 & 2 \\ 1 & -3 \end{pmatrix}$

$B.A = \begin{pmatrix} 3 & 2 \\ 1 & -3 \end{pmatrix}\begin{pmatrix} 1 & 0 \\ 0 & 1 \end{pmatrix} = \begin{pmatrix} 3 & 2 \\ 1 & -3 \end{pmatrix}$

Hence in multiplication the matrix $\begin{pmatrix} 1 & 0 \\ 0 & 1 \end{pmatrix}$ leaves the matrix unchanged. This matrix is called the identity matrix or the unit matrix.

EXERCISE 18.2

1. Form the matrix products, where possible, from the following matrices

$$A = \begin{pmatrix} 3 \\ 1 \end{pmatrix} \quad B = (5 \quad 2) \quad C = \begin{pmatrix} 1 & -1 \\ 0 & 3 \end{pmatrix} \quad D = \begin{pmatrix} 3 & 2 \\ 1 & -2 \end{pmatrix}$$

$$E = \begin{pmatrix} 1 & 1 & 2 \\ -1 & 0 & -2 \end{pmatrix}$$

2. Given that $A = \begin{pmatrix} 2 & 1 \\ 3 & 4 \end{pmatrix}$ $B = \begin{pmatrix} 1 & 0 \\ 2 & 1 \end{pmatrix}$ $C = \begin{pmatrix} 4 & -1 \\ 2 & 0 \end{pmatrix}$

find (a) $A.B$, $B.C$, $A.C$ and $C.A$, (b) A^2, B^2, C^2 (c) B^3, $B^2 - A^2$, $A^2 - B.A$

3. In question 3 of EXERCISE 18.1 a car costs £6500, a van costs £5200 and a motorcycle costs £1200. Find the total value of sales at each garage during the week and the total value of stock remaining at each garage at the end of the week.

18.3 DETERMINANTS

The value of $ad - bc$ can be represented by a determinant which is written within vertical parallel lines

$$\begin{vmatrix} a & b \\ c & d \end{vmatrix} = ad - bc$$

A matrix has a determinant if it is a square matrix. Thus the matrix $\begin{pmatrix} 3 & 2 \\ -1 & 2 \end{pmatrix}$ has a determinant $\begin{vmatrix} 3 & 2 \\ -1 & 2 \end{vmatrix}$

$$= 3 \times 2 - 2 \times (-1)$$
$$= 8$$

Notice that a matrix represents an array of numbers but a determinant represents a number.

Example 18.4 Find the values of the determinants of the following matrices

(a) $\begin{pmatrix} 3 & 2 \\ 4 & 1 \end{pmatrix}$ (b) $\begin{pmatrix} 2 & -1 \\ -3 & 2 \end{pmatrix}$ (c) $\begin{pmatrix} 2 & 1 \\ 4 & 0 \end{pmatrix}$

(a) $\begin{vmatrix} 3 & 2 \\ 4 & 1 \end{vmatrix} = 3 \times 1 - 2 \times 4 = -5$

(b) $\begin{vmatrix} 2 & -1 \\ -3 & 2 \end{vmatrix} = 2 \times 2 - (-1) \times (-3) = 1$

(c) $\begin{vmatrix} 2 & 1 \\ 4 & 0 \end{vmatrix} = 2 \times 0 - 1 \times 4 = -4$

Determinants can be used to solve equations.
To solve $ax + by = p$ (1)
$cx + dy = q$ (2)

Multiply (1) by c and (2) by a and subtract (1) from (2)

$$ady - bcy = aq - pc$$
$$(ad - bc)y = aq - pc$$

$$y = \frac{aq - pc}{ad - bc} = \frac{\begin{vmatrix} a & p \\ c & q \end{vmatrix}}{\begin{vmatrix} a & b \\ c & d \end{vmatrix}}$$

similarly $x = \dfrac{\begin{vmatrix} p & b \\ q & d \end{vmatrix}}{\begin{vmatrix} a & b \\ c & d \end{vmatrix}}$

That is form the determinant of the coefficients of the unknowns, this forms the bottom line for x and y. The top line is this same determinant with the coefficients of the unknown we wish to find replaced by the constants from the right hand side of the equations.

Example 18.5 Solve the equations $3x - 2y = 5$
$$2x + y = 7$$

$$x = \frac{\begin{vmatrix} 5 & -2 \\ 7 & 1 \end{vmatrix}}{\begin{vmatrix} 3 & -2 \\ 2 & 1 \end{vmatrix}} = \frac{19}{7} = 2.71, \quad y = \frac{\begin{vmatrix} 3 & 5 \\ 2 & 7 \end{vmatrix}}{\begin{vmatrix} 3 & -2 \\ 2 & 1 \end{vmatrix}} = \frac{11}{7} = 1.57$$

$$x = 2.71, y = 1.57$$

EXERCISE 18.3

1. Find the values of the determinants of the following matrices
(a) $\begin{pmatrix} 3 & 2 \\ -1 & 4 \end{pmatrix}$ (b) $\begin{pmatrix} -2 & 1 \\ 2 & 2 \end{pmatrix}$ (c) $\begin{pmatrix} 1 & 0 \\ 2 & 3 \end{pmatrix}$ (d) $\begin{pmatrix} 2 & 3 \\ 4 & -1 \end{pmatrix}$

2. Solve the following pairs of simultaneous equations:
(a) $3x - 5y = 13$ (b) $2x - y = 11$ (c) $x - 3y = 7$
$3x + 2y = -1$ $x + 3y = -12$ $2x + y = -5$

18.4 THE INVERSE MATRIX

The inverse matrix of A is defined as the matrix A^{-1} such that $A.A^{-1} = A^{-1}.A = I$ where $I = \begin{pmatrix} 1 & 0 \\ 0 & 1 \end{pmatrix}$

If we consider $A = \begin{pmatrix} a & b \\ c & d \end{pmatrix}$

and its inverse is $A^{-1} = \begin{pmatrix} p & q \\ r & s \end{pmatrix}$

then $\begin{pmatrix} a & b \\ c & d \end{pmatrix}\begin{pmatrix} p & q \\ r & s \end{pmatrix} = \begin{pmatrix} 1 & 0 \\ 0 & 1 \end{pmatrix}$

$$\begin{pmatrix} ap+br & aq+bs \\ cp+dr & cq+ds \end{pmatrix} = \begin{pmatrix} 1 & 0 \\ 0 & 1 \end{pmatrix}$$

We therefore have the equations $ap + br = 1$
$$cp + dr = 0$$

Solving these using determinants we have

$$p = \frac{\begin{vmatrix} 1 & b \\ 0 & d \end{vmatrix}}{\begin{vmatrix} a & b \\ c & d \end{vmatrix}} = \frac{d}{\begin{vmatrix} a & b \\ c & d \end{vmatrix}} \qquad r = \frac{\begin{vmatrix} a & 1 \\ c & 0 \end{vmatrix}}{\begin{vmatrix} a & b \\ c & d \end{vmatrix}} = \frac{-c}{\begin{vmatrix} a & b \\ c & d \end{vmatrix}}$$

Similarly $aq + bs = 0$ gives the solutions
$$cq + ds = 1$$

$$q = \frac{\begin{vmatrix} 0 & b \\ 1 & d \end{vmatrix}}{\begin{vmatrix} a & b \\ c & d \end{vmatrix}} = \frac{-b}{\begin{vmatrix} a & b \\ c & d \end{vmatrix}} \qquad s = \frac{\begin{vmatrix} a & 0 \\ c & 1 \end{vmatrix}}{\begin{vmatrix} a & b \\ c & d \end{vmatrix}} = \frac{a}{\begin{vmatrix} a & b \\ c & d \end{vmatrix}}$$

Hence the inverse matrix is A^{-1} where:

$$A^{-1} = \begin{pmatrix} \dfrac{d}{\begin{vmatrix} a & b \\ c & d \end{vmatrix}} & \dfrac{-b}{\begin{vmatrix} a & b \\ c & d \end{vmatrix}} \\ \dfrac{-c}{\begin{vmatrix} a & b \\ c & d \end{vmatrix}} & \dfrac{a}{\begin{vmatrix} a & b \\ c & d \end{vmatrix}} \end{pmatrix} \quad \text{or } A^{-1} = \frac{1}{\begin{vmatrix} a & b \\ c & d \end{vmatrix}} \begin{pmatrix} d & -b \\ -c & a \end{pmatrix}$$

To find the inverse of a 2×2 matrix we must interchange the terms in the leading diagonal and change the sign of the other terms. We then divide each term by the determinant of the matrix.

Example 18.6 Find the inverses of the following matrices and show that $A.A^{-1} = A^{-1}.A = I$

(a) $\begin{pmatrix} 2 & 1 \\ 5 & 3 \end{pmatrix}$ (b) $\begin{pmatrix} 4 & 2 \\ 5 & 1 \end{pmatrix}$

(a) $\begin{vmatrix} 2 & 1 \\ 5 & 3 \end{vmatrix} = 6 - 5 = 1$,

$$A^{-1} = \frac{1}{1}\begin{pmatrix} 3 & -1 \\ -5 & 2 \end{pmatrix} = \begin{pmatrix} 3 & -1 \\ -5 & 2 \end{pmatrix}$$

$$A.A^{-1} = \begin{pmatrix} 2 & 1 \\ 5 & 3 \end{pmatrix}\begin{pmatrix} 3 & -1 \\ -5 & 2 \end{pmatrix} = \begin{pmatrix} 1 & 0 \\ 0 & 1 \end{pmatrix}$$

$$A^{-1}.A = \begin{pmatrix} 3 & -1 \\ -5 & 2 \end{pmatrix}\begin{pmatrix} 2 & 1 \\ 5 & 3 \end{pmatrix} = \begin{pmatrix} 1 & 0 \\ 0 & 1 \end{pmatrix}$$

(b) $\begin{vmatrix} 4 & 2 \\ 5 & 1 \end{vmatrix} = 4 - 10 = -6, \ A^{-1} = -\frac{1}{6}\begin{pmatrix} 1 & -2 \\ -5 & 4 \end{pmatrix}$

$$A.A^{-1} = \begin{pmatrix} 4 & 2 \\ 5 & 1 \end{pmatrix} \times -\frac{1}{6}\begin{pmatrix} 1 & -2 \\ -5 & 4 \end{pmatrix}$$

$$= -\frac{1}{6}\begin{pmatrix} -6 & 0 \\ 0 & -6 \end{pmatrix} = \begin{pmatrix} 1 & 0 \\ 0 & 1 \end{pmatrix}$$

$$A^{-1}.A = -\frac{1}{6}\begin{pmatrix} 1 & -2 \\ -5 & 4 \end{pmatrix}\begin{pmatrix} 4 & 2 \\ 5 & 1 \end{pmatrix}$$

$$= -\frac{1}{6}\begin{pmatrix} -6 & 0 \\ 0 & -6 \end{pmatrix} = \begin{pmatrix} 1 & 0 \\ 0 & 1 \end{pmatrix}$$

It will be noticed that if the determinant of the matrix is zero then an inverse cannot be found. For example $\begin{pmatrix} 1 & 2 \\ 3 & 6 \end{pmatrix}$ has a determinant of value zero, hence no inverse can be found. Such matrices are called singular matrices. Matrices can be used to solve simultaneous equations but for equations in two unknowns the method is longer than other methods.

Example 18.7 Solve the equations $3x + y = 1$
$$x + 2y = 2$$

These can be written in matrix form as

$$\begin{pmatrix} 3 & 1 \\ 1 & 2 \end{pmatrix}\begin{pmatrix} x \\ y \end{pmatrix} = \begin{pmatrix} 1 \\ 2 \end{pmatrix}$$

We now multiply both sides by the inverse of $\begin{pmatrix} 3 & 1 \\ 1 & 2 \end{pmatrix}$

which is $\frac{1}{5}\begin{pmatrix} 2 & -1 \\ -1 & 3 \end{pmatrix}$

$$\frac{1}{5}\begin{pmatrix} 2 & -1 \\ -1 & 3 \end{pmatrix}\begin{pmatrix} 3 & 1 \\ 1 & 2 \end{pmatrix}\begin{pmatrix} x \\ y \end{pmatrix} = \frac{1}{5}\begin{pmatrix} 2 & -1 \\ -1 & 3 \end{pmatrix}\begin{pmatrix} 1 \\ 2 \end{pmatrix}$$

The LHS always reduces to $\begin{pmatrix} x \\ y \end{pmatrix}$ since $A^{-1}.A = I$

Hence we have $\begin{pmatrix} x \\ y \end{pmatrix} = \frac{1}{5}\begin{pmatrix} 0 \\ 5 \end{pmatrix} = \begin{pmatrix} 0 \\ 1 \end{pmatrix}$

The solution is $x = 0, y = 1$

EXERCISE 18.4

1. Find the inverses of the following matrices and show that $A.A^{-1} = A^{-1}.A = I$ in each case.

(a) $\begin{pmatrix} 3 & 1 \\ 2 & 1 \end{pmatrix}$ (b) $\begin{pmatrix} 4 & 2 \\ 1 & 3 \end{pmatrix}$ (c) $\begin{pmatrix} 4 & -2 \\ 1 & 3 \end{pmatrix}$ (d) $\begin{pmatrix} 3 & -2 \\ -4 & 5 \end{pmatrix}$

2. Which of the following matrices are singular matrices?

(a) $\begin{pmatrix} 2 & 1 \\ 1 & 2 \end{pmatrix}$ (b) $\begin{pmatrix} 3 & 1 \\ 6 & 2 \end{pmatrix}$ (c) $\begin{pmatrix} 3 & 2 \\ 0 & 0 \end{pmatrix}$ (d) $\begin{pmatrix} 1 & 0 \\ 1 & 0 \end{pmatrix}$

(e) $\begin{pmatrix} 3 & 2 \\ 2 & 1 \end{pmatrix}$ (f) $\begin{pmatrix} 3 & 2 \\ 6 & 4 \end{pmatrix}$

3. Solve for x and y the following

(a) $\begin{pmatrix} x+1 \\ y-3 \end{pmatrix} = \begin{pmatrix} 2 \\ 2 \end{pmatrix}$ (b) $\begin{pmatrix} 3x \\ 2y \end{pmatrix} = \begin{pmatrix} 12 \\ 8 \end{pmatrix}$

(c) $\begin{pmatrix} 3x & 2x \\ y & 3y \end{pmatrix}\begin{pmatrix} 2 \\ 1 \end{pmatrix} = \begin{pmatrix} 16 \\ 15 \end{pmatrix}$ (d) $\begin{pmatrix} 1 & 1 \\ 3 & y \end{pmatrix}\begin{pmatrix} x \\ 1 \end{pmatrix} = \begin{pmatrix} 4 \\ 1 \end{pmatrix}$

4. Use a matrix method to solve

(a) $3x + y = -1$ (b) $2x - y = -11$
    $x + 2y = -1$ $x + 3y = -12$

(c) $3x - 5y = 13$
    $3x + 2y = -1$

18.5 GENERAL EXAMPLES

Example 18.8 For the circuit shown in fig. 18.1 calculate the current in each section

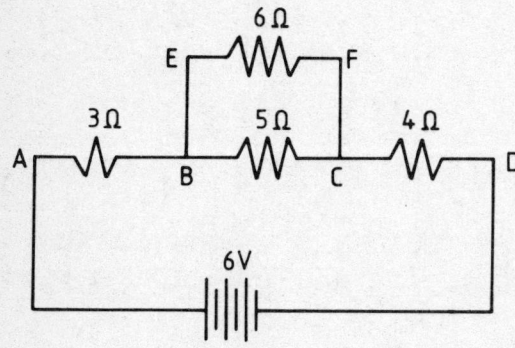

Figure 18.1

Let the current through EF be I_1 amps
Let the current through BC be I_2 amps
Then the current through AB and CD will be $I_1 + I_2$ amps
Consider the circuit ABCD

$$3(I_1 + I_2) + 5I_2 + 4(I_1 + I_2) = 6$$
$$7I_1 + 12I_2 = 6 \qquad (1)$$

Consider the circuit ABEFCD

$$3(I_1 + I_2) + 6I_1 + 4(I_1 + I_2) = 6$$
$$13I_1 + 7I_2 = 6 \qquad (2)$$

Using matrices to solve the equations (1) and (2)

$$\begin{pmatrix} 7 & 12 \\ 13 & 7 \end{pmatrix}\begin{pmatrix} I_1 \\ I_2 \end{pmatrix} = \begin{pmatrix} 6 \\ 6 \end{pmatrix}$$

The inverse of $\begin{pmatrix} 7 & 12 \\ 13 & 7 \end{pmatrix}$ is $-\dfrac{1}{107}\begin{pmatrix} 7 & -12 \\ -13 & 7 \end{pmatrix}$

$$-\frac{1}{107}\begin{pmatrix} 7 & -12 \\ -13 & 7 \end{pmatrix}\begin{pmatrix} 7 & 12 \\ 13 & 7 \end{pmatrix}\begin{pmatrix} I_1 \\ I_2 \end{pmatrix}$$

$$= -\frac{1}{107}\begin{pmatrix} 7 & -12 \\ -13 & 7 \end{pmatrix}\begin{pmatrix} 6 \\ 6 \end{pmatrix}$$

$$\begin{pmatrix} I_1 \\ I_2 \end{pmatrix} = \begin{pmatrix} \dfrac{30}{107} \\ \dfrac{36}{107} \end{pmatrix} = \begin{pmatrix} 0.28 \\ 0.34 \end{pmatrix}$$

Hence $I_1 = 0.28$ A and $I_2 = 0.34$ A

Hence current through AB is 0.62 A, BC is 0.34 A, EF is 0.28 A and CD is 0.62 A.

EXERCISE 18.5

1. Calculate the value of the determinant of the following matrices (a) $\begin{pmatrix} 4 & -3 \\ 2 & 6 \end{pmatrix}$ (b) $\begin{pmatrix} -2 & 3 \\ 2 & 4 \end{pmatrix}$ (c) $\begin{pmatrix} 8 & -6 \\ -2 & 9 \end{pmatrix}$

Solve the following for x and y

(a) $\begin{pmatrix} 1 & 3 \\ 4 & 5 \end{pmatrix}\begin{pmatrix} 3 \\ -2 \end{pmatrix} = \begin{pmatrix} x \\ y \end{pmatrix}$ (b) $\begin{pmatrix} 3 & 4 \\ 5 & 6 \end{pmatrix}\begin{pmatrix} x \\ y \end{pmatrix} = \begin{pmatrix} -5 \\ -7 \end{pmatrix}$

3. Solve the following equations (i) using determinants
                                 (ii) using matrices

(a) $3x + 5y = 19$ (b) $7x - 3y = 12$
    $2x - y = 4$ $4y - 2x = 7$

4. Calculate the current in each section of the circuits given in (a) fig. 18.2, (b) fig. 18.3

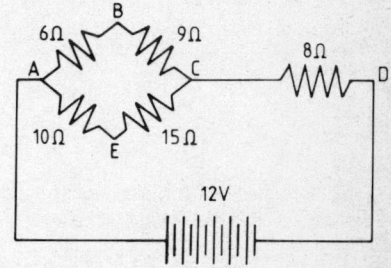

Figure 18.2

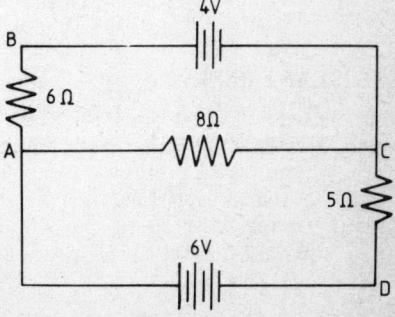

Figure 18.3

SUMMARY

1. A matrix is an array of numbers with m rows and n columns.

2. Two matrices with the same number of rows and columns can be added or subtracted. Corresponding elements are added or subtracted.

3. The product $A.B$ can only be found if A has n columns and B has n rows.

4. When A and B are square matrices $A.B$ and $B.A$ can be found but $A.B$ does not always equal $B.A$.

5. The matrix $I = \begin{pmatrix} 1 & 0 \\ 0 & 1 \end{pmatrix}$ is called the unit matrix.

$A.I = I.A = A$

6. The determinant of $\begin{pmatrix} a & b \\ c & d \end{pmatrix}$ is $\begin{vmatrix} a & b \\ c & d \end{vmatrix} = ad - bc$

7. The equations $ax + by = p$ can be solved by determinants
$cx + dy = q$

$$x = \frac{\begin{vmatrix} p & b \\ q & d \end{vmatrix}}{\begin{vmatrix} a & b \\ c & d \end{vmatrix}}, \quad y = \frac{\begin{vmatrix} a & p \\ c & q \end{vmatrix}}{\begin{vmatrix} a & b \\ c & d \end{vmatrix}}$$

8. The inverse of $A = \begin{pmatrix} a & b \\ c & d \end{pmatrix}$

is $A^{-1} = \dfrac{1}{\begin{vmatrix} a & b \\ c & d \end{vmatrix}} \begin{pmatrix} d & -b \\ -c & a \end{pmatrix}$

and $A.A^{-1} = A^{-1}.A = I$

9. When the determinant of a matrix is zero, the matrix is said to be singular and no inverse exists.

10. The equations $ax + by = p$ can be solved by matrices.
$cx + dy = q$

If $A = \begin{pmatrix} a & b \\ c & d \end{pmatrix}$ then $A^{-1}.A\begin{pmatrix} x \\ y \end{pmatrix} = A^{-1}\begin{pmatrix} p \\ q \end{pmatrix}$

SELF ASSESSMENT PAPER No 18

Instructions: Answer all questions in both sections
Time allowed: Section A 30 minutes (40 marks)
Section B 30 minutes
(20 marks each question)
Marks gained: 40+ pass with credit, 32–40 pass, less than 32 fail, repeat chapter 18

Section A

$A = \begin{pmatrix} 2 & 3 \\ -1 & 4 \end{pmatrix}$ and $B = \begin{pmatrix} 6 & 2 \\ 4 & 1 \end{pmatrix}$

1. Find (a) $A + B$, (b) $A - B$

2. Find (a) $A.B$, (b) $B.A$

3. Find the determinant of (a) A, (b) B

4. Find the inverse of A

5. Find the value of $2A + I$ where I is the unit matrix.

6. Which of the following matrices are singular?

(a) $\begin{pmatrix} 3 & 2 \\ 6 & 4 \end{pmatrix}$ (b) $\begin{pmatrix} 1 & 6 \\ 0 & 1 \end{pmatrix}$ (c) $\begin{pmatrix} 0 & 3 \\ 0 & 4 \end{pmatrix}$ (d) $\begin{pmatrix} 2 & 2 \\ 2 & 0 \end{pmatrix}$

Section B

1. Use determinants to solve the equations
$3x + 7y = 15$
$2x - 5y = 7$

2. Use matrices to solve the equations $4I_1 - 3I_2 = 9$
$2I_1 + 7I_2 = 17$

ANSWERS

Exercise 18.1   1. $A + D = \begin{pmatrix} 6 & 3 \\ 5 & 7 \end{pmatrix}$, $B + G = \begin{pmatrix} 6 \\ 7 \end{pmatrix}$

$E + F = (5 \quad 8)$

2. (a) $\begin{pmatrix} 2 & 5 \\ 1 & 3 \end{pmatrix}$, (b) $\begin{pmatrix} 4 & -1 \\ 9 & -1 \end{pmatrix}$, (c) $\begin{pmatrix} 1 & 2 \\ -7 & 6 \end{pmatrix}$

(d) $\begin{pmatrix} 1 & 3 \\ 8 & -3 \end{pmatrix}$, (e) $\begin{pmatrix} 6 & 4 \\ 10 & .2 \end{pmatrix}$, (f) $\begin{pmatrix} 6 & -3 \\ -9 & 12 \end{pmatrix}$,

(g) $\begin{pmatrix} 9 & -5 \\ 22 & -4 \end{pmatrix}$, (h) $\begin{pmatrix} 8 & -14 \\ 10 & 0 \end{pmatrix}$,

3. $\begin{pmatrix} 7 & 3 & 2 \\ 8 & 5 & 4 \\ 11 & 3 & 5 \end{pmatrix}, \begin{pmatrix} 3 & 1 & 1 \\ 3 & 3 & 2 \\ 5 & 0 & 2 \end{pmatrix}, \begin{pmatrix} 4 & 2 & 1 \\ 5 & 2 & 2 \\ 6 & 3 & 3 \end{pmatrix}$

Exercise 18.2   1. $A.B = \begin{pmatrix} 15 & 6 \\ 5 & 2 \end{pmatrix}$, $B.C = (5 \quad 1)$,

$B.D = (17 \quad 6)$

$B.E = (3 \quad 5 \quad 6)$, $C.A = \begin{pmatrix} 2 \\ 3 \end{pmatrix}$, $C.D = \begin{pmatrix} 2 & 4 \\ 3 & -6 \end{pmatrix}$,

$C.E = \begin{pmatrix} 2 & 1 & 4 \\ -3 & 0 & -6 \end{pmatrix}$, $D.A = \begin{pmatrix} 11 \\ 1 \end{pmatrix}$, $D.C = \begin{pmatrix} 3 & 3 \\ 1 & -7 \end{pmatrix}$,

$D.E = \begin{pmatrix} 1 & 3 & 2 \\ 3 & 1 & 6 \end{pmatrix}$

2. (a) $\begin{pmatrix} 4 & 1 \\ 11 & 4 \end{pmatrix} \begin{pmatrix} 4 & -1 \\ 10 & -2 \end{pmatrix}, \begin{pmatrix} 10 & -2 \\ 20 & -3 \end{pmatrix}, \begin{pmatrix} 5 & 0 \\ 4 & 2 \end{pmatrix}$

(b) $\begin{pmatrix} 7 & 6 \\ 18 & 19 \end{pmatrix}, \begin{pmatrix} 1 & 0 \\ 4 & 1 \end{pmatrix}, \begin{pmatrix} 14 & -4 \\ 8 & -2 \end{pmatrix},$ (c) $\begin{pmatrix} 1 & 0 \\ 6 & 1 \end{pmatrix}$

$\begin{pmatrix} -6 & -6 \\ -14 & -18 \end{pmatrix}, \begin{pmatrix} 5 & 5 \\ 11 & 13 \end{pmatrix},$

3. $\begin{pmatrix} 25\,900 \\ 37\,500 \\ 34\,900 \end{pmatrix}, \begin{pmatrix} 37\,600 \\ 45\,300 \\ 58\,200 \end{pmatrix}$

Exercise 18.3   1. (a) 14, (b) -6, (c) 3, (d) -14,
2.   (a) $x = 1, y = -2$, (b) $x = 3, y = -5$, (c) $x = -1\frac{1}{7}$,
$y = -2\frac{5}{7}$

Exercise 18.4   1. (a) $\begin{pmatrix} 1 & -1 \\ -2 & 3 \end{pmatrix}$, (b) $\begin{pmatrix} 0.3 & -0.2 \\ -0.1 & 0.4 \end{pmatrix}$

(c) $\frac{1}{14}\begin{pmatrix} 3 & 2 \\ -1 & 4 \end{pmatrix}$, (d) $\frac{1}{7}\begin{pmatrix} 5 & 2 \\ 4 & 3 \end{pmatrix}$,   2. (b), (c), (d), (f),

3. (a) $x = 1, y = 5$, (b) $x = 4, y = 4$, (c) $x = 2, y = 3$,
(d) $x = 3, y = -8$,   4. (a) $x = -\frac{1}{5}, y -\frac{2}{5}$,
(b) $x = -6\frac{3}{7}, y = -1\frac{6}{7}$, (c) $x = 1, y = -2$

Exercise 18.5   1. (a) 30, (b) -14, (c) 60,
2. (a) $x = -3, y = 2$, (b) $x = 1, y = -2$,
3. (a) $x = 3, y = 2$, (b) $x = \dfrac{69}{22}$ (3.136), $y = \dfrac{72}{22}$ (3.318)
4. (a) ABC 0.432 A, AEC 0.258 A, CD 0.690 A,
(b) AB 0.033 A, CD 0.441 A, AC 0.474 A.

SELF ASSESSMENT PAPER No 18

Marks

Section A

1. (a) $\begin{pmatrix} 8 & 5 \\ 3 & 5 \end{pmatrix}$ 4

   (b) $\begin{pmatrix} -4 & 1 \\ -5 & 3 \end{pmatrix}$ 4

2. (a) $\begin{pmatrix} 24 & 7 \\ 10 & 2 \end{pmatrix}$ 4

   (b) $\begin{pmatrix} 10 & 10 \\ 7 & 16 \end{pmatrix}$ 4

3. (a) 11   3

   (b) -2 3

4. Determinant $= 11$ 4

   $\frac{1}{11}\begin{pmatrix} 4 & -3 \\ 1 & 2 \end{pmatrix}$ 4

5. $2A = \begin{pmatrix} 4 & 6 \\ -2 & 8 \end{pmatrix}$ 2

$= \begin{pmatrix} 5 & 6 \\ -2 & 9 \end{pmatrix}$ **Marks**   4

6. (a), (c) are singular matrices   4

Section B

1. $x = \dfrac{\begin{vmatrix} 15 & 7 \\ 7 & -5 \end{vmatrix}}{\begin{vmatrix} 3 & 7 \\ 2 & -5 \end{vmatrix}}$ 4

$x = \dfrac{-75 - 49}{-15 - 14}$ 4

$x = 4.28$ 2

$y = \dfrac{\begin{vmatrix} 3 & 15 \\ 2 & 7 \end{vmatrix}}{-29}$ 4

$y = \dfrac{21 - 30}{-29}$ 4

$y = 0.31$ 2

2. $A^{-1} = \dfrac{1}{34}\begin{pmatrix} 7 & 3 \\ -2 & 4 \end{pmatrix}$ 8

$\begin{pmatrix} I_1 \\ I_2 \end{pmatrix} = \dfrac{1}{34}\begin{pmatrix} 7 & 3 \\ -2 & 4 \end{pmatrix}\begin{pmatrix} 9 \\ 17 \end{pmatrix}$ 4

$\begin{pmatrix} I_1 \\ I_2 \end{pmatrix} = \dfrac{1}{34}\begin{pmatrix} 114 \\ 50 \end{pmatrix}$ 4

$I_1 = 3.35$ 2

$I_2 = 1.47$ 2

PHASE TEST 6

Instructions: Answer all questions in Section A and two questions in Section B
Time allowed: Section A 20 minutes (20 marks)
Section B 40 minutes
(20 marks each question)
Marks gained: 30+ pass with credit, 24–30 pass, less than 24 fail, repeat chapters 17 and 18

Section A

1. State what variables should be plotted to give a straight line graph for the following laws
(a) $y = a\sqrt{x} + b$, (b) $y = at + bt^2$, (c) $y = ax^n$.
2. Sketch the curves identified by (a) $y = mx + c$, (b) $y = ax^2 + bx + c$, (c) $y = \dfrac{c}{x}$. Name each type of curve.

Marks

3. $A = \begin{pmatrix} 3 & -1 \\ 2 & 2 \end{pmatrix}$, $B = \begin{pmatrix} 2 & 1 \\ -1 & 3 \end{pmatrix}$

Find (a) $A + B$, (b) $A.B$, (c) the determinant of A, (d) the inverse of A.

Section B

1. By drawing a suitable graph show that the following results indicate a relationship between x and y of the form $y = \dfrac{a}{x^2} + b$ and determine the values of a and b.

x	10	15	20	25	30	35
y	4	-1.6	-3.5	-4.4	-4.89	-5.18

2. It is thought that a law of the form $Q = aH^n$ connects the quantities Q and H. Draw a suitable straight line graph to illustrate this and hence determine the values of the constants a and n.

H	2.2	1.8	1.4	1.1	0.8	0.6
Q	8.9	8.03	7.23	6.4	5.5	4.85

3. Use matrices to solve the equations

$$I_1 + 2I_2 = 4$$
$$-3I_1 + I_2 = -3.25$$

ANSWERS PHASE TEST 6

Marks

Section A

1. (a) Plot y against $\sqrt{x}$ — 2
   (b) Plot $\dfrac{y}{t}$ against t — 2
   (c) Plot $\log y$ against $\log x$ — 2

2. (a) Sketch of straight line — 2
   (b) Sketch of parabola — 2
   (c) Sketch of rectangular hyperbola — 2

3. (a) $A + B = \begin{pmatrix} 5 & 0 \\ 1 & 5 \end{pmatrix}$ — 2

   (b) $A.B = \begin{pmatrix} 7 & 0 \\ 2 & 8 \end{pmatrix}$ — 2

   (c) $\begin{vmatrix} 3 & -1 \\ 2 & 2 \end{vmatrix} = 6 + 2 = 8$ — 2

(d) Inverse of $A = \frac{1}{8}\begin{pmatrix} 2 & 1 \\ -2 & 3 \end{pmatrix}$ — 2

Section B

1. Plot y against $\dfrac{1}{x^2}$ — 2

y	4	-1.6	-3.5	-4.4	-4.89	-5.18
$\dfrac{1}{x^2}$	0.0100	0.0044	0.0025	0.0016	0.0011	0.0008

— 6

Graph of y vertically against $\dfrac{1}{x^2}$ — 5

Intercept on y-axis, $b = -6$ — 3
Gradient $a = 1000$ — 4

2. Plot $\log_{10} Q$ against $\log_{10} H$ — 2
   $(\log Q = n \log H + \log a)$

$\log_{10} H$	0.342	0.255	0.146	0.041	-0.097	-0.222
$\log_{10} Q$	0.949	0.905	0.859	0.806	0.740	0.685

— 6

Graph of $\log_{10} Q$ vertically against $\log H$ — 5
Take two points from the graph
$\log_{10} Q = 0.7$, $\log_{10} H = -0.18$;
$0.7 = -0.18n + \log_{10} a$ .... (1)
$\log_{10} Q = 0.9$, $\log_{10} H = 0.25$;
$0.9 = 0.25n + \log_{10} a$ .... (2) — 2
Take (1) from (2) $0.2 = 0.43n$, $n = 0.47$ — 2
Substitute into (2) $0.9 = 0.25 \times 0.43 + \log_{10} a$ — 1
$\log_{10} a = 0.7825$, $a = 6.06$ — 2

3. $\begin{pmatrix} 1 & 2 \\ -3 & 1 \end{pmatrix}\begin{pmatrix} I_1 \\ I_2 \end{pmatrix} = \begin{pmatrix} 4 \\ -3.25 \end{pmatrix}$ — 2

Inverse $= \dfrac{1}{7}\begin{pmatrix} 1 & -2 \\ 3 & 1 \end{pmatrix}$ — 7

$\begin{pmatrix} I_1 \\ I_2 \end{pmatrix} = \dfrac{1}{7}\begin{pmatrix} 1 & -2 \\ 3 & 1 \end{pmatrix}\begin{pmatrix} 4 \\ -3.25 \end{pmatrix}$ — 4

$\begin{pmatrix} I_1 \\ I_2 \end{pmatrix} = \dfrac{1}{7}\begin{pmatrix} 10.5 \\ 8.75 \end{pmatrix}$ — 4

$\begin{pmatrix} I_1 \\ I_2 \end{pmatrix} = \begin{pmatrix} 1.5 \\ 1.25 \end{pmatrix}$ — 2

$I_1 = 1.5$, $I_2 = 1.25$ — 1

19 Differentiation

In this chapter the derivative of y with respect to x, $\frac{dy}{dx}$, is developed and the method of determining this quantity is explained. Since $\frac{dy}{dx}$ gives the rate of change of y as x changes it is a very important function with wide applications. Some of its immediate uses are velocity $\frac{ds}{dt}$, the rate of change of distance with time; acceleration $\frac{dv}{dt}$, the rate of change of velocity with time and electric current $\frac{dq}{dt}$, the rate of change of charge, q, with time. Many more examples will be met in advanced electrical and mechanical engineering.

19.1 DIFFERENTIATION FROM FIRST PRINCIPLES

One quantity is said to be a function of another when its value depends on the value of the second quantity.

Example 19.1 If $y = x^2 + 2x + 3$ find the values of y when $x = 1, 2, 0, a + b$,
Since $y = x^2 + 2x + 3$ then y is a function of x and is written $y = f(x)$.

Thus $f(x) = x^2 + 2x + 3$ for all values of x

when $x = 1$; $f(1) = 1^2 + 2 \times 1 + 3 = 6$

when $x = 2$; $f(2) = 2^2 + 2 \times 2 + 3 = 11$

when $x = 0$; $f(0) = 0^2 + 2 \times 0 + 3 = 3$

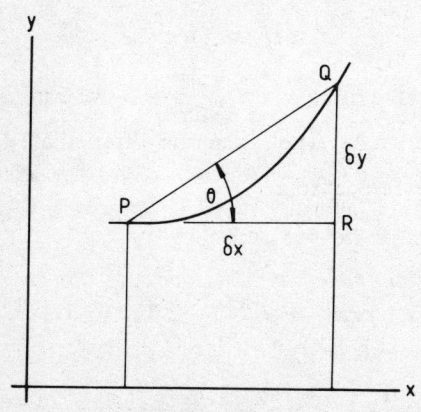

Figure 19.1

when $x = a + b$; $f(a + b) = (a + b)^2 + 2(a + b) + 3$

We will now use δx and δy. δx means 'a little bit of x' and δy means 'a little bit of y'. The δ must not be separated from the x or y.

Consider the curve $y = f(x)$, shown in fig. 19.1

If P is the point (x, y) on the curve and Q is a point near to P and also on the curve then Q is said to be the point

$$(x + \delta x, \ y + \delta y)$$

or $\qquad y + \delta y = f(x + \delta x)$

The gradient of PQ is the increase in y divided by the increase in x. Gradient of PQ $= \dfrac{QR}{PR} = \dfrac{\delta y}{\delta x}$

The actual gradient of the curve at P is the tangent of the curve at the point P. Now as Q moves nearer and nearer to P so the gradient of PQ will become nearer and nearer to the gradient of the curve at P. Thus the gradient of the curve at P is $\dfrac{\delta y}{\delta x}$ when δx is very small.

We say that the gradient of the curve at P is the limit of $\dfrac{\delta y}{\delta x}$ as δx approaches zero.

$$\underset{\delta x \to 0}{\text{Lt}} \ \frac{\delta y}{\delta x} = \text{gradient of the curve at P}$$

Example 19.2 Consider the curve $y = x^2$ when $x = 2$. We have $y = x^2$ for P and $y + \delta y = (x + \delta x)^2$ for Q

δx	$x + \delta x$	y	$y + \delta y$	δy	$\dfrac{\delta y}{\delta x}$
1	3	4	9	5	5
0.1	2.1	4	4.41	0.41	4.1
0.01	2.01	4	4.0401	0.0401	4.01
0.001	2.001	4	4.004 001	0.004 001	4.001

Therefore as x becomes smaller and smaller the value of $\dfrac{\delta y}{\delta x}$ becomes nearer and nearer to 4.

At the point $x = 2$ on the curve $y = x^2$, $\underset{\delta x \to 0}{\text{Lt}} \ \dfrac{\delta y}{\delta x} = 4$

We call this the first derivative of y with respect to x

and it is written $\dfrac{dy}{dx} = \underset{\delta x \to 0}{\text{Lt}} \ \dfrac{\delta y}{\delta x}$

$\dfrac{dy}{dx}$ is the rate of change of y per unit change of x.

In the above it is the gradient of the curve.

Example 19.3 Find the value of $\dfrac{dy}{dx}$ on the curve $y = x^2 + 2x + 3$ where $x = 1$.

Since the point $(x + \delta x, y + \delta y)$ lies on the curve

$$y + \delta y = (x + \delta x)^2 + 2(x + \delta x) + 3$$

$$y + \delta y = x^2 + 2x\delta x + (\delta x)^2 + 2x + 2\delta x + 3$$

We now subtract $y = x^2 + 2x + 3$ from this

$$y + \delta y - y = x^2 + 2x\delta x + (\delta x)^2 + 2x + 2\delta x$$
$$+ 3 - x^2 - 2x - 3$$
$$y = 2x\delta x + (\delta x)^2 + 2\delta x$$

Divide both sides by δx

$$\frac{\delta y}{\delta x} = 2x + \delta x + 2$$

Now $\qquad \dfrac{dy}{dx} = \underset{\delta x \to 0}{\text{Lt}} \dfrac{\delta y}{\delta x} = 2x + 2$

We can ignore the δx since this is very small compared with the $2x + 2$.

$$\frac{dy}{dx} = 2x + 2$$

When $x = 1$, $\dfrac{dy}{dx} = 2 \times 1 + 2 = 4$

Example 19.4 Find $\dfrac{dy}{dx}$ when $y = 2x^3 + \dfrac{1}{x}$

Working as in the previous example.

$$y + \delta y = 2(x + \delta x)^3 + \frac{1}{x + \delta x}$$

$$y + \delta y = 2x^3 + 6x^2\delta x + 6x(\delta x)^2 + 2(\delta x)^3 + \frac{1}{x + \delta x}$$

Subtract $y = 2x^3 + \dfrac{1}{x}$

$$\delta y = 6x^2\delta x + 6x(\delta x)^2 + 2(\delta x)^3 + \frac{1}{x + \delta x} - \frac{1}{x}$$

$$\delta y = 6x^2\delta x + 6x(\delta x)^2 + 2(\delta x)^3 + \frac{x - (x + \delta x)}{x(x + \delta x)}$$

$$\delta y = 6x^2\delta x + 6x(\delta x)^2 + 2(\delta x)^3 - \frac{\delta x}{x(x + \delta x)}$$

Dividing by δx

$$\frac{\delta y}{\delta x} = 6x^2 + 6x\delta x + 2(\delta x)^2 - \frac{1}{x(x + \delta x)}$$

Now as $\delta x \to 0$, $6x\delta x \to 0$, $2(\delta x)^2 \to 0$, $x(x + \delta x) \to x(x) \to x^2$

Hence $\dfrac{dy}{dx} = \underset{\delta x \to 0}{\text{Lt}} \dfrac{\delta y}{\delta x} = 6x^2 - \dfrac{1}{x^2}$

EXERCISE 19.1

1. Evaluate the following functions when $x = 0, 1, -1$
   (a) $f(x) = x^3 + 6x - 2$, (b) $f(x) = (x - 1)(2x + 3)$,
   (c) $f(x) = x + \log_{10}(2x + 3)$

2. Evaluate the following functions when $x = 0$, $\dfrac{\pi}{3}$ and $\dfrac{\pi}{2}$
   (a) $f(x) = 3 + \sin x$, (b) $f(x) = \tan x + 5 \cos x$

3. Differentiate the following from first principles
   (a) $y = 3x + 1$, (b) $y = 2x^2 + 3x + 1$, (c) $y = x^3 - 2$,
   (d) $y = x + \dfrac{1}{x}$, (e) $y = x(x - 1)$, (f) $y = \dfrac{x - 1}{x^2}$

19.2 DIFFERENTIATING BY RULE

The method of obtaining $\dfrac{dy}{dx}$ in the last pargraph is obviously rather long, hence the following rule is used. If a is a constant then we have found from the above that when

$$y = ax^3, \qquad \frac{dy}{dx} = 3ax^2$$

$$y = ax^2, \qquad \frac{dy}{dx} = 2ax$$

$$y = ax, \qquad \frac{dy}{dx} = a \quad (ax^0)$$

$$y = a, \qquad \frac{dy}{dx} = 0 \quad (0.ax^{-1})$$

$$y = \frac{a}{x}, \qquad \frac{dy}{dx} = -\frac{a}{x^2} \quad (-ax^{-2})$$

$$y = \frac{a}{x^2}, \qquad \frac{dy}{dx} = -\frac{2a}{x^3} \quad (-2ax^{-3})$$

This suggests that when $y = ax^n$, $\dfrac{dy}{dx} = nax^{n-1}$

This is true for all values of n.

Example 19.5 Differentiate the following:

(a) $x^3 - 2x^2 + 3$, (b) $x^{2.5} - x^{1.2} + x^{-0.1}$, (c) $\dfrac{1}{\sqrt{x}} + 3$,
(d) $\dfrac{x^2 + 3x + 2}{x}$, (e) $(x - 1)(x + 2)$

Instead of writing $y = x^2$, $\dfrac{dy}{dx} = 2x$ we can write this as $\dfrac{d}{dx}(x^2) = 2x$. The $\dfrac{d}{dx}$ means differentiate the function with respect to x.

(a) $\dfrac{d}{dx}(x^3 - 2x^2 + 3) = 3x^2 - 4x$

(b) $\dfrac{d}{dx}(x^{2.5} - x^{1.2} + x^{-0.1}) = 2.5x^{1.5} - 1.2x^{0.2}$
$\qquad\qquad - 0.1x^{-1.1}$

(c) $\dfrac{d}{dx}\left(\dfrac{1}{\sqrt{x}} + 3\right) = \dfrac{d}{dx}(x^{-1/2} + 3) = -\tfrac{1}{2}x^{-3/2}$

$$= -\frac{1}{2\sqrt{x^3}}$$

(d) $\frac{d}{dx}\left(\frac{x^2 + 3x + 2}{x}\right) = \frac{d}{dx}(x + 3 + 2x^{-1})$

$$= 1 + 0 - 2x^{-2} = 1 - \frac{2}{x^2}$$

(e) $\frac{d}{dx}([x - 1][x + 2]) = \frac{d}{dx}(x^2 + x - 2) = 2x + 1$

We notice that when we have found $\frac{dy}{dx}$ it is a function of x and hence we can differentiate the function again, to obtain the second derivative $\frac{d}{dx}\left(\frac{dy}{dx}\right) = \frac{d^2 y}{dx^2}$

Example 19.6 Find the first and second derivatives $\frac{dy}{dx}$ and $\frac{d^2 y}{dx^2}$ when $y = 3x^4 + 5x^2 + 2$. Also find the value of these derivatives when $x = 1$ and when $x = -2$.

$$y = 3x^4 + 5x^2 + 2, \quad \frac{dy}{dx} = 12x^3 + 10x,$$

$$\frac{d^2 y}{dx^2} = 36x^2 + 10$$

When $x = 1$, $\frac{dy}{dx} = 12 \times 1^3 + 10 \times 1 = 22$,

$$\frac{d^2 y}{dx^2} = 36 \times 1^2 + 10 = 46$$

If $x = -2$, $\frac{dy}{dx} = 12 \times (-2)^3 + 10 \times (-2) = -116$,

$$\frac{d^2 y}{dx^2} = 36 \times (-2)^2 + 10 = 154$$

EXERCISE 19.2

1. Differentiate the following: (a) $y = 7x^3 + 2x + 1$,

(b) $y = x^2 - \frac{2}{x^2}$, (c) $y = 5\sqrt{x} + 6$, (d) $y = 5t^2 - 2\sqrt{t}$,

(e) $\theta = t^2 - \frac{5}{t}$, (f) $y = 6x^{0.2} - 5x^{1.2}$,

2. Evaluate the following: (a) $\frac{d}{dx}(6x^2 - 2x + 5)$ when $x = 2$, (b) $\frac{d}{dt}(5t^{-0.1} + 6t^{1.3})$ when $t = 3$, (c) $\frac{d}{dx}\left(x^5 - \frac{5}{x^5}\right)$ when $x = 1$, (d) $\frac{d}{d\theta}(\theta^2 - \sqrt{\theta^3})$ when $\theta = 9$.

3. Find $\frac{dy}{dx}$ and $\frac{d^2 y}{dx^2}$ when (a) $y = x^5 - x^3 + 6x$,

(b) $y = \frac{1}{x} - 3\sqrt{x}$, (c) $y = x^{0.2} + x^{1.5}$

4. If $y = 2x^2 + 3$ show that $x\frac{d^2 y}{dx^2} - \frac{dy}{dx} = 0$

5. If $x = 1 - 2\sqrt{t}$ show that $4t^2 \frac{d^2 x}{dt^2} + x = 1$

19.3 DIFFERENTIATION OF sin x AND cos x

Draw the graph of $y = \sin x$, see fig. 19.2, and measure the gradient of the curve at various points on the curve.

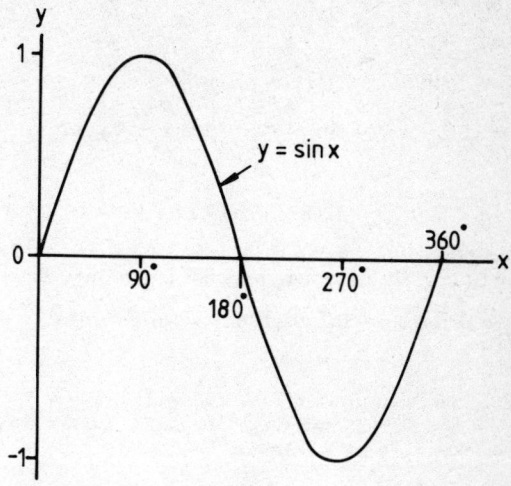

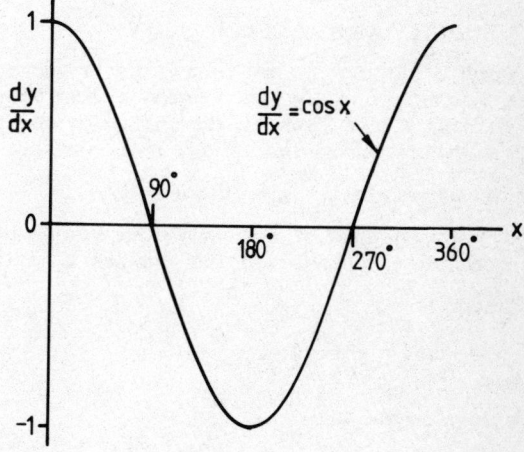

Figure 19.2

The gradient of the curve at $x = 0$ is 1.
The gradient of the curve then decreases to zero when $x = 90°$. The gradient is then negative and becomes larger and larger negatively until at $180°$ it is -1. It then becomes smaller and smaller, negatively, to zero at $270°$ etc.

The gradient is $\frac{dy}{dx}$, and if we draw the graph of $\frac{dy}{dx}$

under the graph of $y = \sin x$ we find we have the curve of $\cos x$.

Hence we have $\dfrac{dy}{dx} = \cos x$

so that $\dfrac{d}{dx}(\sin x) = \cos x$

If we do the same with $y = \cos x$ we will get the curve of $-\sin x$, so that $\dfrac{d}{dx}(\cos x) = -\sin x$

Example 19.7 Find $\dfrac{dy}{dx}$ when (a) $y = 2\sin x$, (b) $y = 3\cos x - 4\sin x$

(a) $y = 2\sin x$, $\dfrac{dy}{dx} = 2\cos x$

(b) $\dfrac{d}{dx}(3\cos x - 4\sin x) = -3\sin x - 4\cos x$

EXERCISE 19.3

1. Draw the graph of $y = \cos x$ from $x = 0°$ to $x = 360°$. Measure the gradients at various points and so plot the curve for $\dfrac{dy}{dx}$. Derive from the graph that when $y = \cos x$, $\dfrac{dy}{dx} = -\sin x$.

2. Find the derivatives of the following: (a) $y = 3\cos x$, (b) $y = 4\sin x$, (c) $y = 5\sin x - 2\cos x$, (d) $y = \cos x - 2\sin x$, (e) $y = -4\cos x - 2\sin x$.

3. When $y = 2\sin x - \cos x$ show that $\dfrac{d^2y}{dx^2} + y = 0$

19.4 VELOCITY AND ACCELERATION

If a graph of distance against time is drawn then the gradient at any point gives the velocity at that point. The gradient is also given by the first derivative of distance with respect to time. Hence if the distance is s and the time is t then $\dfrac{ds}{dt}$ gives the velocity.

Similarly the gradient of a velocity–time graph gives the acceleration. Therefore if the velocity is v, the acceleration is $\dfrac{dv}{dt}$.

But $\dfrac{dv}{dt} = \dfrac{d}{dt}(v) = \dfrac{d}{dt}\left(\dfrac{ds}{dt}\right) = \dfrac{d^2s}{dt^2}$

Hence the acceleration is $\dfrac{dv}{dt}$ or $\dfrac{d^2s}{dt^2}$

Example 19.8 The distance s metres moved by a body in time t seconds is given by $s = t^3 - 2t^2 + 3t - 7$. Find the speed and acceleration at the end of 5 s. When is the acceleration zero?

$s = t^3 - 2t^2 + 3t - 7$, $\dfrac{ds}{dt} = 3t^2 - 4t + 3$, $\dfrac{d^2s}{dt^2} = 6t - 4$

At the end of 5 s the speed, $\dfrac{ds}{dt} = 3 \times 5^2 - 4 \times 5 + 3 = 58$ m/s

At the end of 5 s the acceleration, $\dfrac{d^2s}{dt^2} = 6 \times 5 - 4 = 26$ m/s^2

When the acceleration is zero, $\dfrac{d^2s}{dt^2} = 0$

Hence $6t - 4 = 0$ or $t = \tfrac{2}{3}$ s

The acceleration is zero at the end of $\tfrac{2}{3}$ s

Example 19.9 A wheel rotates through θ radians in t seconds according to the law $\theta = 15 + 10t - t^2$. Find (a) the angular velocity at the end of 3 s, (b) the initial angular velocity, (c) the angular acceleration, (d) the time taken for the wheel to come to rest.

Angular velocity, $\omega = \dfrac{d\theta}{dt} = 10 - 2t$ rad/s

Angular acceleration, $\alpha = \dfrac{d\omega}{dt} = -2$ rad/s^2

(a) At the end of 3 s, $\omega = 10 - 2 \times 3 = 4$ rad/s

(b) The initial value of $\omega = 10 - 2 \times 0 = 10$ rad/s

(c) The angular acceleration is constant, $\alpha = -2$ rad/s^2

(d) When the wheel comes to rest $\omega = 0$

$$10 - 2t = 0 \text{ or } t = 5 \text{ s}$$

EXERCISE 19.4

1. The distance s metres moved by a body in time t seconds is given by $s = 7 + 3t + 6t^2 - t^3$. Find (a) its speed at the end of 3 s, (b) its acceleration at the end of 3 s, (c) the time taken for its speed to reach zero, (d) the time taken for its acceleration to reach zero.

2. A wheel rotates through θ radians in t seconds and θ is given by $\theta = 9 - t^3$. Find, (a) the initial angular velocity and acceleration, (b) the angular velocity and acceleration at the end of 5 s, (c) at what time the wheel will come to rest.

3. The distance s metres moved by a body in t seconds is given by $s = t^3 + 3t^2 + 4$. Find the velocity and acceleration after 3 s.

4. After t seconds, the velocity, v metres per second, of a point moving in a straight line is given by $v = 4t^2 + 5t$. Find the acceleration after 4 s.

19.5 MAXIMA AND MINIMA

A point such as A, in fig. 19.3 is called a maximum point, that is y has a value, at A, greater than its value immediately either side of A. At a point such as A it can be seen that $\dfrac{dy}{dx} = 0$ and $\dfrac{d^2y}{dx^2}$ is negative.

Similarly a point such as B is called a minimum point since values of y either side of B are greater than the value of y at B.

At a point such as B $\dfrac{dy}{dx} = 0$ and $\dfrac{d^2y}{dx^2}$ is positive.

Note: It is the maximum and minimum values of y that are being considered, the values of x indicate the

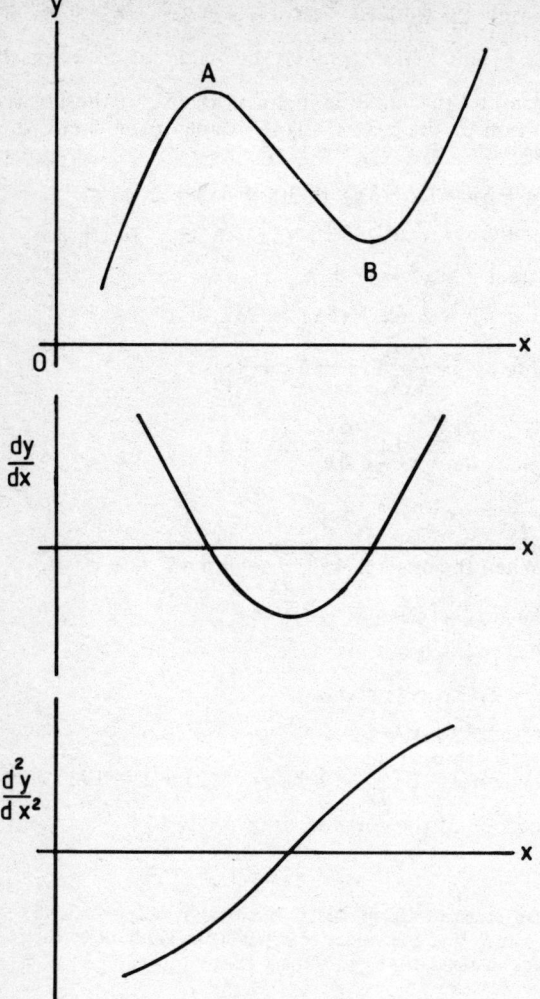

Figure 19.3

positions at which these maximum and minimum values occur.

For a maximum value of y: $\dfrac{dy}{dx} = 0$, $\dfrac{d^2y}{dx^2}$ is $-$ve.

For a minimum value of y: $\dfrac{dy}{dx} = 0$, $\dfrac{d^2y}{dx^2}$ is $+$ve.

These maximum and minimum values are often referred to as stationary values. They are the values where the rate of change is zero.

Example 19.10 Find the stationary values of the following and discriminate between them: (a) $y = x^2 - 6x + 6$, (b) $y = 2x^3 - 3x^2 - 12x + 18$.

(a) $\dfrac{dy}{dx} = 2x - 6$, $\dfrac{d^2y}{dx^2} = 2$

For stationary values $\dfrac{dy}{dx} = 0$, $2x - 6 = 0$ or $x = 3$

When $x = 3$, $\dfrac{d^2y}{dx^2} = 2$ hence the value of y is a minimum.

When $x = 3$, $y = -4$

$y = x^2 - 6x + 6$ has a minimum value of -4 at $x = 3$.

(b) $\dfrac{dy}{dx} = 6x^2 - 6x - 12$ and $\dfrac{d^2y}{dx^2} = 12x - 6$

For stationary values $\dfrac{dy}{dx} = 0$ $6x^2 - 6x - 12 = 0$

$$x^2 - x - 2 = 0$$
$$(x - 2)(x + 1) = 0$$
$$x = 2 \text{ or } x = -1$$

When $x = 2$, $\dfrac{d^2y}{dx^2} = 18$ and therefore $y = -2$ is a minimum value.

When $x = -1$, $\dfrac{d^2y}{dx^2} = -18$ and therefore $y = 25$ is a maximum value.

$y = 2x^3 - 3x^2 - 12x + 18$ has a maximum value of 25 at $x = -1$ and a minimum value of -2 at $x = 2$.

Example 19.11 A cylindrical tin can is made from very thin sheet metal and has a close fitting lid which is bent at the perimeter to overlap the curved surface of the can by 2.5 mm. The total area of the metal used in the construction is $100\pi\,\text{mm}^2$. Prove that if the radius is r millimetres then the volume $V\,\text{mm}^3$ of the can is given by $V = \pi(50r - r^3 - 2.5r^2)$, and find the dimensions of the can for maximum volume.
A diagram of the can is shown in fig. 19.4.

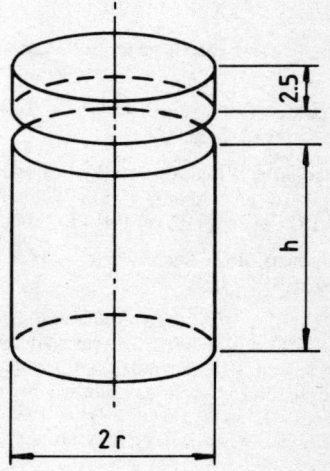

Figure 19.4

Area of metal used = area of lid + area of tin.

$$100\pi = (\pi r^2 + 2\pi r \times 2.5) + (\pi r^2 + 2\pi r h)$$
$$100\pi = \pi r^2 + 5\pi r + \pi r^2 + 2\pi r h$$
$$100\pi = 2\pi r^2 + 5\pi r + 2\pi r h$$

$2\pi rh = 100\pi - 2\pi r^2 - 5\pi r$

$$h = \frac{100\pi - 2\pi r^2 - 5\pi r}{2\pi r}$$

The volume of the can $V = \pi r^2 h$

$$V = \pi r^2 \left(\frac{100\pi - 2\pi r^2 - 5\pi r}{2\pi r} \right)$$

$$V = r \left(\frac{100\pi - 2\pi r^2 - 5\pi r}{2} \right)$$

$V = \pi(50r - r^3 - 2.5r^2)$

$\dfrac{dV}{dr} = \pi(50 - 3r^2 - 5r), \quad \dfrac{d^2V}{dr^2} = \pi(0 - 6r - 5)$

For maximum volume $\dfrac{dV}{dr} = 0$, $\pi(50 - 3r^2 - 5r) = 0$

$50 - 3r^2 - 5r = 0$

$3r^2 + 5r - 50 = 0$

$(3r - 10)(r + 5) = 0$

$3r - 10 = 0$, $r = \frac{10}{3}$ or $r + 5 = 0$, $r = -5$

We must ignore $r = -5$ since the radius cannot be negative

When $r = \frac{10}{3}$, $h = \dfrac{100\pi - 2\pi(\frac{10}{3})^2 - 5\pi \times \frac{10}{3}}{2\pi \times \frac{10}{3}} = \frac{55}{6}$

$V = \pi r^2 h = \pi(\frac{10}{3})^2(\frac{55}{6}) = 320 \text{ mm}^3$

For maximum volume $r = 3\frac{1}{3}$ cm, $h = 9\frac{1}{6}$ cm and the maximum volume is 320 mm^3.

EXERCISE 19.5

1. Find the gradient of the curve $y = x^3 + 2x + 5$ at the point where $x = 1$.

2. Find the point on the curve $y = x^2 + 3x + 2$ where the curve is parallel to the x-axis.

3. Find the stationary values on the following curves and discriminate between them: (a) $y = 2 - 9x + 6x^2 - x^3$, (b) $y = 2x^3 - 15x^2 + 36x + 6$, (c) $y = 12 - 24x - 9x^2 + 2x^3$.

4. Find the gradient of the curve $y = x^2 - 5x + 6$ at the points where it intersects the x-axis.

5. (a) Show that the area of an equilateral triangle of side x is $0.433x^2$. (b) The perimeter of a sheet of metal is 1 m and it consists of an equilateral triangle with a rectangle on one side. If the equilateral triangle has side x show that the area of the metal sheet is $0.5x - 1.067x^2$. Hence find the dimensions of the sheet of metal when it has a maximum area.

6. A piece of sheet metal consists of a semicircle radius r with a rectangle of length L on its diameter.
(a) If the area is to be $10\,000 \text{ mm}^2$ prove that the perimeter is $\dfrac{10\,000}{r} + r\left(\dfrac{\pi}{2} + 2\right)$ and find the radius r and length L in order that the perimeter shall be a minimum.
(b) Calculate the minimum perimeter.

19.6 GENERAL EXAMPLES

Example 19.12 (a) If $y = x^2 + 3x + 2$ find from first principles the value of $\dfrac{dy}{dx}$. (b) If $y = 3x^3 - 6x^2 + 5x + 7$ find $\dfrac{dy}{dx}$ and hence find (i) the value of x when the tangent to the curve is inclined at $45°$ to the positive direction of the x-axis. (ii) the slope of the curve when $x = 2$.

(a) $y + \delta y = (x + \delta x)^2 + 3(x + \delta x) + 2$

$\quad y + \delta y = x^2 + 2x\delta x + (\delta x)^2 + 3x + 3\delta x + 2$

Subtract $y = x^2 + 3x + 2$

$\qquad \delta y = 2x\delta x + (\delta x)^2 + 3\delta x$

Divide by δx $\quad \dfrac{\delta y}{\delta x} = 2x + \delta x + 3$

$$\frac{dy}{dx} = \operatorname*{Lt}_{\delta x \to 0} \frac{\delta y}{\delta x} = 2x + 3$$

(b) $\dfrac{dy}{dx} = 9x^2 - 12x + 5$

(i) When the angle is $45°$, $\dfrac{dy}{dx} = \tan 45° = 1$

$9x^2 - 12x + 5 = 1$

$9x^2 - 12x + 4 = 0$

$(3x - 2)(3x - 2) = 0$

$3x - 2 = 0$, $x = \frac{2}{3}$

(ii) When $x = 2$, $\dfrac{dy}{dx} = 9(2)^2 - 12(2) + 5 = 17$

The slope of the curve when $x = 2$ is 17

EXERCISE 19.6

1. Differentiate from first principles $y = 2x^3 + x^2 - 5x + 3$ and hence find the values of x when the gradient to the curve makes an angle of $45°$ with the x-axis.

2. (a) Differentiate (i) $y = \dfrac{2x^3 + 5x - 1}{x^2}$, (ii) $y = 6\sqrt{x} = \sqrt{x^3}$, (iii) $y = (3 - x)^2$.
(b) Find for what values of x the gradient of the curve $y = 2x^3 - 15x^2 + 42x + 17$ has the value 6.

3. (a) Find whether the function $x^2 - 3x + 5$ has a maximum or minimum value.
(b) Find for what values of x the gradient of the curve $y = 2x^3 + 15x^2 + 12x - 17$ has the value 48.

4. A body is thrown vertically upwards so that its height h metres at t seconds is given by the relation $h = 100t - 4t^2$. Find the greatest height reached.

5. (a) Differentiate with respect to x: $3x^{2/3} + x^{-1/2} + 5x^{1.2} + 4$.
(b) A flywheel rotates according to the law $\theta = 28t - 3t^2$, θ giving the angular rotation in radians taking place in t seconds. Find expressions for the angular velocity $\dfrac{d\theta}{dt}$ and the angular acceleration $\dfrac{d^2\theta}{dt^2}$. What is the value of t when $\dfrac{d\theta}{dt} = 0$? Find the value of θ for this value of t.

6. Given that $I = 2 + 3t^2$ and $V = 5I + \dfrac{dI}{dt}$, find the positive

value of t which makes $V = 20$.

SUMMARY

1. One quantity is said to be a function of another when its value depends on the value of the second quantity e.g. $f(x) = x^2 - 6x + 7$.

2. The first derivative $\dfrac{dy}{dx} = \underset{\delta x \to 0}{\text{Lt}} \dfrac{\delta y}{\delta x}$. The second derivative is represented by $\dfrac{d^2 y}{dx^2}$.

3. When $y = ax^n$, $\dfrac{dy}{dx} = nax^{n-1}$ for all values of n.

4. When $y = \sin x$, $\dfrac{dy}{dx} = \cos x$; when $y = \cos x$, $\dfrac{dy}{dx} = -\sin x$.

5. When the distance s metres moved by a body in time t seconds is given by $s = f(t)$, $\dfrac{ds}{dt} =$ the velocity, $\dfrac{d^2 s}{dt^2} = \dfrac{dy}{dt} =$ the acceleration.

6. When $y = f(x)$; a maximum value of y occurs when $\dfrac{dy}{dx} = 0$ and $\dfrac{d^2 y}{dx^2}$ is $-$ ve; a minimum value of y occurs where $\dfrac{dy}{dx} = 0$ and $\dfrac{d^2 y}{dx^2}$ is $+$ ve.

SELF ASSESSMENT PAPER No 19

Instructions: Answer all questions in both sections
Time allowed: Section A 30 minutes (40 marks)
Section B 30 minutes (20 marks each question)
Marks gained: 40+ pass with credit, 32–40 pass, less than 32 fail, repeat Chapter 19.

Section A

1. Differentiate from first principles $y = x^2 - 2x + 6$

2. Differentiate the following: (a) $y = 3x^3 - 2x^2 + 7$, (b) $y = (2 - x)(x + 5)$.

3. Find $\dfrac{d}{dx}\left(x^3 - \dfrac{1}{x^2}\right)$ when $x = 2$.

4. Find $\dfrac{dy}{dx}$ and $\dfrac{d^2 y}{dx^2}$ when $y = x^2 - 2\sqrt{x}$

5. Find $\dfrac{dy}{dx}$ when $y = 3 \cos x - 4 \sin x$.

6. Find the maximum or minimum value of $y = x^2 - 4x + 6$, stating whether it is a maximum or minimum value.

Section B

1. Differentiate the following with respect to x:

(a) $\dfrac{x^3 + 2x^2 + 1}{x^2}$, (b) $x(x^2 + 1)(x - 2)$, (c) $(x + 3)^2$,
(d) $5 \sin x - 2 \cos x$.

2. A body moves a distance s metres in t seconds, given by $s = t^4 - 6t^3 - 24t^2 + 16t - 7$
(i) Find an expression for the velocity, v, in terms of t.
(ii) Find $\dfrac{dv}{dt}$ and $\dfrac{d^2 v}{dt^2}$ and hence find the value of t when the velocity is a minimum.
(iii) Calculate this minimum velocity.
(iv) Calculate the distance moved when the velocity is a minimum.

ANSWERS

Exercise 19.1   1.(a) -2, 5, -9, (b) -3, 0, -2, (c) 0.4771, 1.6990, -1,   2.(a) 3, 3.866, 4. (b) 5, 4.232, ∞,   3.(a) 3, (b) $4x + 3$, (c) $3x^2$, (d) $1 - \dfrac{1}{x^2}$, (e) $2x - 1$, (f) $\dfrac{2 - x}{x^3}$

Exercise 19.2   1.(a) $21x^2 + 2$, (b) $2x + \dfrac{4}{x^3}$, (c) $\dfrac{5}{2\sqrt{x}}$, (d) $10t - \dfrac{1}{\sqrt{t}}$,   (e) $2t + \dfrac{5}{t^2}$,   (f) $1.2x^{-0.8} - 6x^{0.2}$, 2.(a) 22,   (b) 10.75,   (c) 30,   (d) 13.5,   3.(a) $5x^4 - 3x^2 + 6$, $20x^3 - 6x$,   (b)$\dfrac{1}{-x^2} - \dfrac{3}{2\sqrt{x}}$, $\dfrac{2}{x^3} + \dfrac{3}{4\sqrt{x^3}}$, (c) $0.2x^{-0.8} + 1.5x^{0.5}$, $-0.16x^{-1.8} + 0.75x^{-0.5}$

Exercise 19.3   2.(a) $-3 \sin x$, (b) $4 \cos x$, (c) $5 \cos x + 2 \sin x$, (d) $-\sin x - 2 \cos x$, (e) $4 \sin x - 2 \cos x$.

Exercise 19.4   1.(a) 12 m/s, (b) -6 m/s^2, (c) 3.732 s, 0.268 s, (d) 2 s,   2.(a) $0, 0$, (b) -75 rad/s, -30 rad/s^2, (c) 0,   3. 45 m/s, 24 m/s^2,   4. 37 m/s^2.

Exercise 19.5   1. 5,   2. $x = -\tfrac{3}{2}, y = \tfrac{1}{4}$, 3.(a) $x = 3$, $y = 2$ max, $x = 1$, $y = -2$ min, (b) $x = 3$, $y = 33$ min, $x = 2$, $y = 34$ max, (c) $x = 4$, $y = -100$ min, $x = -1$, $y = 25$ max,   4. $-1, 1$, 5. radians $= 0.234$ m, length $= 0.149$ m, 6. $r = 52.9$ mm, $L = 53.4$ mm, perimeter $= 379$ mm

Exercise 19.6   1. $6x^2 + 2x - 5$, $x = -1.1805$, $x = 0.8471$,   2.(a) (i) $2 - \dfrac{5}{x^2} + \dfrac{2}{x^3}$, (ii) $\dfrac{3}{\sqrt{x}} - \dfrac{3\sqrt{x}}{2}$, (iii) $2x - 6$,   (b) $2, 3$,   3.(a) min, (b) $-6, 1$, 4. 625 m,   5. (a) $\dfrac{2}{\sqrt[3]{x}} - \dfrac{1}{2\sqrt{x^3}} + 6x^{0.2}$, (b) $t = 4\tfrac{2}{3}$, $\theta = 65\tfrac{1}{3}$,   6. 0.641 s

SELF ASSESSMENT PAPER No 19

Marks

Section A

1. $y + \delta y = (x + \delta x)^2 - 2(x + \delta x) + 6$ $\qquad$ 2

$\delta y = 2x\delta x + (\delta x)^2 - 2\delta x$ 2

$\dfrac{\delta y}{\delta x} = 2x + \delta x - 2$ 2

$\dfrac{dy}{dx} = 2x - 2$ 2

2. (a) $9x^2 - 4x$ 2

    (b) $10 - 3x - x^2$ 2

      $\dfrac{dy}{dx} = -3 - 2x$ 2

3. $\dfrac{d}{dx}(x^3 - x^{-2})$ 2

    $= 3x^2 + \dfrac{2}{x^3}$ 3

    $= 12\frac{1}{4}$ 2

4. $y = x^2 - 2x^{1/2}$ 2

   $\dfrac{dy}{dx} = 2x - x^{-1/2} = 2x - \dfrac{1}{\sqrt{x}}$ 2

   $\dfrac{d^2y}{dx^2} = 2 + \frac{1}{2}x^{-3/2} = 2 + \dfrac{1}{2\sqrt{x^3}}$ 2

5. $-3\sin x - 4\cos x$ 4

6. $\dfrac{dy}{dx} = 2x - 4$ 2

   $\dfrac{d^2y}{dx^2} = 2$ min value of y 2

For min $2x - 4 = 0$, $x = 2$ 3
Min value of y is 2   2

Section B

1. (a) $\dfrac{d}{dx}(x + 2 + x^{-2})$ 3

     $= 1 - 2x^{-3} = 1 - \dfrac{2}{x^3}$ 3

   (b) $\dfrac{d}{dx}(x^4 - 2x^3 + x^2 - 2x)$ 3

     $= 4x^3 - 6x^2 + 2x - 2$ 3

   (c) $\dfrac{d}{dx}(x^2 + 6x + 9)$ 2

     $= 2x + 6$ 2

   (d) $5\cos x + 2\sin x$ 4

2. (i) $v = \dfrac{ds}{dt} = 4t^3 - 18t^2 - 48t + 16$ 2

   (ii) $\dfrac{dv}{dt} = 12t^2 - 36t - 48$ 2

     $\dfrac{d^2v}{dt^2} = 24t - 36$ 2

For minimum values $\dfrac{dv}{dt} = 0$ 1

$12 - 36t - 48 = 0$ 1
$t = 4$ or -1 4

When $t = 4$, $\dfrac{d^2v}{dt^2} = 60$, min   2

When $t = -1$, $\dfrac{d^2v}{dt^2} = -60$, max   2

(iii) When $t = 4$, $v = -208$ m/s   2
(iv) When $t = 4$, $s = -455$ m   2

20 The Exponential Function

20.1 THE DERIVATIVE OF e^x

The exponential function is written e^x and is such that its derivative is also e^x.

$$\frac{d}{dx}(e^x) = e^x$$

This means that at any point on the curve $y = e^x$ the gradient of the curve is equal to the y value.

$$e^x = 1 + x + \frac{x^2}{2!} + \frac{x^3}{3!} + \frac{x^4}{4!} + \ldots$$

Where 2! means the factorial 2 and is equal to 2×1
3! means factorial 3 and is equal to $3 \times 2 \times 1$ etc.
We can see that if we differentiate this we have

$$\frac{d}{dx}(e^x) = \frac{d}{dx}\left(1 + x + \frac{x^2}{2!} + \frac{x^3}{3!} + \frac{x^4}{4!} + \ldots\right)$$

$$= 0 + 1 + \frac{2x}{2 \times 1} + \frac{3x^2}{3 \times 2 \times 1}$$

$$+ \frac{4x^3}{4 \times 3 \times 2 \times 1} + \ldots$$

$$= 1 + x + \frac{x^2}{2!} + \frac{x^3}{3!} + \ldots$$

$$\frac{d}{dx}(e^x) = e^x$$

Example 20.1 Draw the graph of $y = e^{2x}$ from $x = -2$ to $x = 1$

x	-2	-1	0	0.5	1
$y = e^{2x}$	0.018	0.135	1	2.718	7.389

The above values can be found from a calculator or from tables.
The graph is drawn in fig. 20.1
If we now measure the gradients at the above points we have

x	-2	-1	0	0.5	1
Gradient of e^{2x}	0.04	0.27	2	5.4	15

We therefore see that the gradients are twice the values of e^{2x}. This suggests that $\frac{d}{dx}(e^{2x}) = 2e^{2x}$
This is true and $\frac{d}{dx}(e^{ax}) = ae^{ax}$ for all values of a.

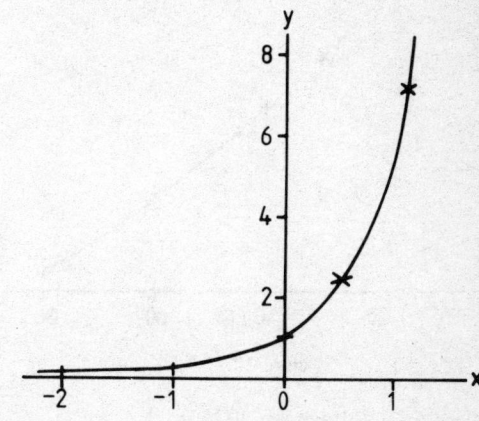

Figure 20.1

This can also be found by differentiating the series for e^{ax}.

$$e^{ax} = 1 + ax + \frac{a^2 x^2}{2!} + \frac{a^3 x^3}{3!} + \frac{a^4 x^4}{4!} + \ldots$$

$$\frac{d}{dx}(e^{ax}) = 0 + a + \frac{2a^2 x}{2 \times 1} + \frac{3a^3 x^2}{3 \times 2 \times 1} + \frac{4a^4 x^3}{4 \times 3 \times 2 \times 1} \ldots$$

$$= a + a^2 x + \frac{a^3 x^2}{2!} + \frac{a^4 x^3}{3!} + \ldots$$

$$= a\left(1 + ax + \frac{a^2 x^2}{2!} + \frac{a^3 x^3}{3!} + \ldots\right)$$

$$\frac{d}{dx}(e^{ax}) = ae^{ax}$$

Example 20.2 Find the derivatives of (a) e^{3x} (b) $e^{-1/2x}$ (c) $3e^{2x}$

(a) $\dfrac{d}{dx}(e^{3x}) = 3e^{3x}$

(b) $\dfrac{d}{dx}(e^{-1/2x}) = -\tfrac{1}{2}e^{-1/2x}$

(c) $\dfrac{d}{dx}(3e^{2x}) = 3 \times 2e^{2x} = 6e^{2x}$

Example 20.3 The charge on a capacitor is falling and the voltage V at any time t seconds is given by $V = V_0 e^{-kt}$, where $V_0 = 200$ and $k = 0.008$. Draw a graph of V against t and determine when the voltage is 60% of its initial value.

t	0	10	20	30	40	50	60	80
V	200	185	170	157	145	134	124	105

The graph is drawn in fig. 20.2

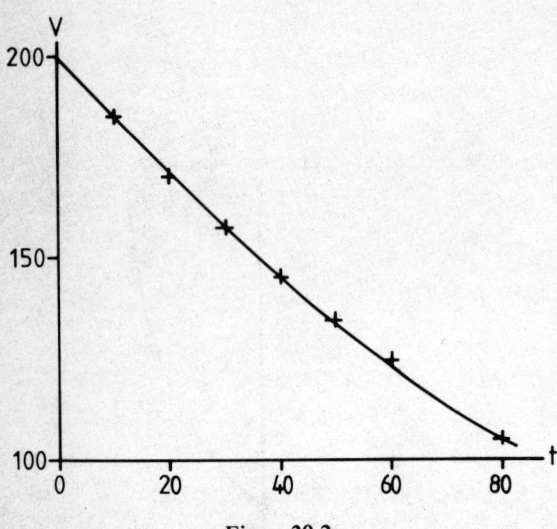

Figure 20.2

60% of the initial voltage is 120 V.
When $V = 120$, $t = 64$ s.
The voltage is 60% of its initial value after 64 s.

EXERCISE 20.1

1. Draw the graph of $y = e^x$, for values of x from -2 to $+2$, and show that the gradient at any point is equal to the value of the function at that point.

2. Draw the graph of $y = e^{-1/2x}$, for values of x from -2 to $+2$, and show that the gradient at any point is equal to $-\frac{1}{2}$ times the value of the function at that point.

3. Find the derivatives of (a) e^{2x}, (b) e^{-3x}, (c) $e^x + e^{-x}$, (d) $5e^{-1/2x}$, (e) $4e^{2x} - 5e^{-x}$.

4. The value of a machine, £P, after time t years is given by $P = Ae^{-kt}$ where A is the initial cost of the machine and k is a constant 0.3. Draw a graph of P against t for a machine that cost £2000 and for values of t from 0 to 10. When will the machine have a value of £800?

5. The current I amperes through a circuit is given by $I = I_0 e^{-kt}$ where $I_0 = 20$ A and $k = 0.25$. Draw a graph of I against t seconds for values of t from 0 to 10 s. When will the current be 30% of its initial value?

20.2 LAWS OF TYPE $y = ae^{kx}$

Since e is an important quantity and frequently occurs in practical problems, tables and calculators normally include the values for e^x and also $ln\,x\,(\log_e)$. Now with the equation $y = ae^{kx}$ if we take logarithms of both sides to base e we have

$ln\,y = ln\,ae^{kx}$

$ln\,y = ln\,a + ln\,e^{kx}$

but $ln\,e^{kx} = kx\,ln\,e = kx$ since $ln\,e = 1$

Hence $ln\,y = kx + ln\,a$

If we plot $ln\,y$ against x we will get a straight line of gradient k and intercept on the $ln\,y$ axis of $ln\,a$.

Example 20.4 Verify that the following values of force T and angle of contact θ, obtained from a polishing process, are connected by the law $T = ae^{\mu\theta}$, and find values of the constants a and μ.

θ (radians)	0.4	0.7	1.0	1.3	1.6
T	18.2	19.8	21.9	23.9	26.3

Taking logarithms to base e of both sides

$ln\,T = ln\,ae^{\mu\theta}$

$ln\,T = ln\,a + ln\,e^{\mu\theta}$

$ln\,T = \mu\theta + ln\,a$ (1)

$ln\,T$	2.901	2.991	3.087	3.174	3.270
θ	0.4	0.7	1.0	1.3	1.6

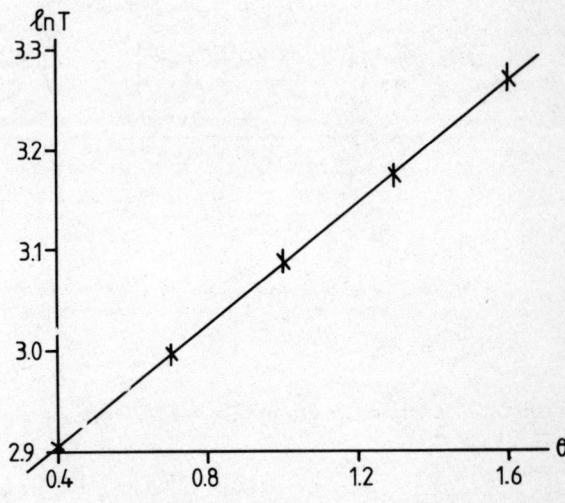

Figure 20.3

Since a straight line can reasonably be drawn through the plotted points the law is verified. We select two points from the graph and substitute them in equation (1) above.

$\theta = 0.6$, $ln\,T = 2.961$ $2.961 = 0.6\mu + ln\,a$ (2)

$\theta = 1.5$, $ln\,T = 3.248$ $3.248 = 1.5\mu + ln\,a$ (3)

subtract (2) from (3) $3.248 - 2.961 = 1.5\mu - 0.6\mu$

$0.287 = 0.9\mu$

$\mu = 0.319$

From (2) when $\mu = 0.319$; $2.961 = 0.6 \times 0.319 + \ln a$

$\ln a = 2.961 - 0.6 \times 0.319$

$\qquad = 2.770$

$\qquad a = $ antilog of $2.770 = 16.0$

Note that the antilog of $\ln a$ is $e^{2.77}$ since if $e^y = a$ then $y = \ln a$.
The law is $T = 16.0e^{0.319\theta}$

EXERCISE 20.2

1. Examine whether the following experimental values of x and y appear to follow a law of the form $y = ke^{cx}$, and if so find the most suitable values of k and c.

x	1	2	3	4	5
y	2.00	2.15	2.26	2.43	2.58

2. The current through a circuit is thought to be given by the equation $I = I_0 e^{-kt}$ where k and I_0 are constants and t is the time in seconds. The current was measured at various times and the following values, which may contain experimental error, were obtained:

t(s)	0.45	0.91	2.78	3.40	5.56	8.20
I(A)	18.50	13.20	9.80	7.64	3.54	1.76

Verify that the above law holds and find an average value for k and I_0.

20.3 GENERAL EXAMPLES

Example 20.5 (a) Solve the equation $e^{-2x} = 0.6201$
(b) If T_1 and T_2 are the tensions in a belt which passes round a pulley then $\dfrac{T_1}{T_2} = e^{\mu\theta}$ where $\mu = $ coefficient of friction and θ is the angle of contact of the belt and pulley, measured in radians. Calculate T_2 if $T_1 = 200$ N, $\mu = 0.3$ and $\theta = \dfrac{5\pi}{4}$ radians.

(a) $e^{-2x} = 0.6201$

Take logarithms to base e of both sides.

$\ln e^{-2x} = \ln 0.6201$

$-2x \ln e = \ln 0.6201$ but $\ln e = 1$

$-2x = \ln 0.6201 = -0.4779$

$x = 0.2390$

(Note if tables are used to find $\ln 0.6201$ then this is written

$\ln 0.6201 = \ln\left(\dfrac{6.201}{10}\right)$

$\qquad = \ln 6.201 - \ln 10 = 1.8247 - 2.3026)$

(b) $\dfrac{T_1}{T_2} = e^{\mu\theta}$, $\quad \dfrac{200}{T_2} = e^{0.3 \times \frac{5\pi}{4}}$

$T_2 = \dfrac{200}{e^{0.3 \times \frac{5\pi}{4}}} = \dfrac{200}{e^{1.178}} = \dfrac{200}{3.248}$

$T_2 = 61.6$ N

EXERCISE 20.3

1. (a) Solve the equations (i) $e^x = 107.6$, (ii) $e^{2x} = 17$. (b) In the electrical formula $I = I_0 e^{-(Rt/L)}$, $I_0 = 500$, $t = 0.015$, $R = 400$ and $L = 10$. Calculate the value of I.

2. (a) Solve the equations (i) $e^{2x+1} = 43$, (ii) $e^{x^2} = 31$.
(b) From the formula $I = \left(\dfrac{E}{R}\right)(1 - e^{-[Rt/L]})$ calculate the value of I when $R = 800$, $L = 6.3$, $E = 5 \times 10^4$ and $t = 0.018$.

3. The voltage V of a capacitor at time t seconds was measured and recorded below

t	0	10	20	30	40	50
V	400	370	340	310	290	270

Draw a straight line graph to show that these are connected by a law $V = V_0 e^{-kt}$ where V_0 and k are constants. From the graph determine values for the constants V_0 and k.

SUMMARY

1. The exponential function e^x is such that $\dfrac{d}{dx}(e^x) = e^x$
When $y = e^{ax}$ then $\dfrac{dy}{dx} \doteq ae^{ax}$

2. When the graph of $y = e^{ax}$ is plotted the gradient is proportional to the function at any point.

3. The constants in the law $y = ae^{kx}$ can be found by plotting a straight line graph of $\ln y$ against x. The gradient of the straight line is k and the intercept on the $\ln y$ axis is $\ln a$.

SELF ASSESSMENT PAPER No 20

Instructions: Answer both questions
Time allowed: One hour (25 marks each question)
Marks gained: 25+ pass with credit, 20–25 pass, less than 20 fail, repeat Chapter 20.

1. (a) Draw the graph of $y = e^x$ for values of x from -2 to $+2$. Measure the gradient of the graph at $x = -2, -1, 0, 1, 2$ and hence establish the relationship between the gradient of the function and the function itself.
(b) Find the derivatives of (i) e^{2x}, (ii) e^{-5x}, (iii) $3e^x - e^{-x}$.

2. The value £P of a machine after t years service is given in the following table:

t	1	2	3	4	5
P	502	380	287	219	166

Verify graphically that these values satisfy approximately a law of the type $P = ae^{-kt}$ where a and k are constants and find the values of these constants.
How many years will elapse before the value falls below £70?

ANSWERS

Exercise 20.1   3. (a) $2e^{2x}$, (b) $-3e^{-3x}$, (c) $e^x - e^{-x}$, (d) $-2.5e^{-1/2x}$, (e) $8e^{2x} + 5e^{-x}$,
4.   After 3.05 years,   5. $t = 4.82\,\text{s}$.

Exercise 20.2   1. $k = 1.88$, $c = 0.065$,
2. $k = 0.324$, $I_0 = 23.0$

Exercise 20.3   1. (a) (i) 4.678, (ii) 1.417, (b) 274.4,
2. (a) (i) 1.3806, (ii) 1.853, (b) 56,
3. $V_0 = 400$, $k = 0.008$

SELF ASSESSMENT PAPER No 20

Marks

1. (a)

x	-2	-1	0	1	2	
e^x	0.135	0.368	1	2.718	7.389	5

Graph   7

x	-2	-1	0	1	2	
Gradient	0.14	0.40	1	2.7	7.4	5

(b) (i) $2e^{2x}$ 2
     (ii) $-5e^{-5x}$ 2
     (iii) $3e^x + e^{-x}$ 4

2. Plot $ln\,P$ against t 1

t	1	2	3	4	5	
$ln\,P$	6.219	5.940	5.659	5.389	5.112	5

Graph   6
Graph is a straight line hence law is verified   1
Intercept $ln\,a = 6.5$ 3
$a = 665$ 2
Gradient $= -0.278$, $-k = -0.278$ 2
$k = 0.278$ 1
$P = 665e^{-0.278t}$ 1
When $P = 70$, $t = 8$ years   3

21 Integration

21.1 THE INDEFINITE INTEGRAL

In chapter 19 we saw that if $y = x^3$ then $\frac{dy}{dx} = 3x^2$.

In the reverse process if $\frac{dy}{dx} = 3x^2$ then $y = x^3$. This is called integration and is represented by the symbol $\int$.

This symbol is an 's' and stands for summation, in future work on integration it will be seen that integration can also be defined as a summation.

The above can be written $\frac{d}{dx}(y) = 3x^2$ or $d(y) = 3x^2 dx$

and expressing this in the integral notation

$$y = \int 3x^2 dx = x^3$$

But $\frac{d}{dx}(x^3 + 2) = 3x^2$, $\frac{d}{dx}(x^3 + 7) = 3x^2$

and in general $\frac{d}{dx}(x^3 + c) = 3x^2$ where c is any

constant. Therefore $\int 3x^2 dx = x^3 + c$ where c is an

arbitary constant, called the constant of integration. Later it will be shown that the constant of integration can be found if some condition is given to the integral. Integrals in which c is not determined are called indefinite integrals.

In general $\int x^n dx = \frac{x^{n+1}}{n+1} + c$

But it will be seen that this will not work if $n = -1$.

When $n = -1$ $\int \frac{1}{x} dx = \ln x + c$

Example 21.1 Integrate the following with respect to x.

(a) $x^3 + 3x$, (b) $x^{2.3} + x^{1.2} - \frac{1}{x^2}$, (c) $6 - \sqrt{x}$,

(d) $3 \sin x - \cos x$

(a) $\int (x^3 + 3x) dx = \frac{x^4}{x} + \frac{3x^2}{2} + c$

(b) $\int \left(x^{2.3} + x^{1.2} - \frac{1}{x^2} \right) dx = \int (x^{2.3} + x^{1.2} - x^{-2}) dx$

$$= \frac{x^{3.3}}{3.3} + \frac{x^{2.2}}{2.2} - \frac{x^{-1}}{-1} + c$$

$$= 0.303x^{3.3} + 0.455x^{2.2} + \frac{1}{x} + c$$

(c) $\int (6 - \sqrt{x}) dx = \int (6 - x^{1/2}) dx$

$$= 6x - \frac{x^{3/2}}{3/2} + c$$

$$= 6x - \frac{2}{3} \sqrt{x^3} + c$$

(d) $\int (3 \sin x - \cos x) dx = -3 \cos x - \sin x + c$

Note: In the above 6 is treated as $6x^0$ for integrating. Also the integral of $\sin x$ is $-\cos x$ and the integral of $\cos x$ is $\sin x$.

EXERCISE 21.1

Evaluate the following integrals

1. $\int (7x^3 - 2x^2 + 5) dx$

2. $\int (x^2 - 2x + 4) dx$

3. $\int \frac{(x^2 - 1)(x^2 + 1)}{x^2} dx$

4. $\int \left(2 - \frac{x^2}{5} - \frac{1}{\sqrt{x}} \right) dx$

5. $\int \left(\frac{2}{\sqrt{x}} - x \right) dx$

6. $\int (x^{1.2} + x^{2.4}) dx$

7. $\int (\sqrt{x^3} - 7x^2) dx$

8. $\int \left(\frac{3x^5 - 4x^3}{x^5} \right) dx$

9. $\int (3x - 2)(x + 1) dx$

10. $\int \left(x^3 - 6 + \frac{8}{x^3} \right) dx$

11. $\int (2 \sin x + 3 \cos x) dx$

12. $\int (\cos x - \sin x) dx$

21.2 DEFINITE INTEGRALS AND AREAS

Consider the curve $y = f(x)$ where $f(x)$ is the derivative of $F(x)$ or $\int f(x) dx = F(x) + c$ $\qquad$ (1)

The curve is sketched in fig. 21.1.

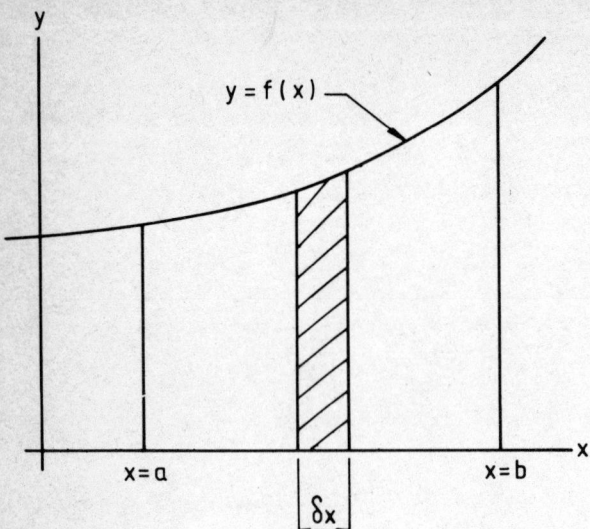

Figure 21.1

Let the shaded element have area δA then $y\delta x < \delta A < (y + \delta y)\delta x$

Divide by δx $y < \dfrac{\delta A}{\delta x} < y + \delta y$

Now in the limit as $\delta x \to 0, \delta y \to 0$ and $\dfrac{\delta A}{\delta x} \to \dfrac{dA}{dx}$

Hence $\dfrac{dA}{dx} = y$

This suggests that $A = \int y\,dx$ and since $y = f(x)$

$$A = \int f(x)\,dx = F(x) + c \quad \text{from (1)}$$

Suppose the area A lies between the ordinates $x = a$ and $x = b$ when $x = a$ the area A is zero

$$0 = F(a) + c \ \text{ or } \ c = -F(a)$$

$$A = F(x) - F(a)$$

When $x = b$ the required area $A = F(b) - F(a)$.
The area between the curve $y = f(x)$, the x-axis and the ordinates $x = a$ and $x = b$ is $F(b) - F(a)$ and this is written

$$\text{Area} = \int_a^b f(x)\,dx$$

a and b are called the limits of the integral and since it contains no constant of integration c it is called a definite integral, that is it has a definite value.

Example 21.2 Find the area between the curve $y = x^2 + 3x + 1$, the x-axis and the ordinates $x = 1$ and $x = 3$, see fig. 21.2

$$\text{Area} = \int_1^3 (x^2 + 3x + 1)\,dx$$

$$= \left[\frac{x^3}{3} + \frac{3x^2}{2} + x\right]_1^3$$

$$= \left(\frac{3^3}{3} + \frac{3 \times 3^2}{2} + 3\right) - \left(\frac{1^3}{3} + \frac{3 \times 1^2}{2} + 1\right)$$

$$= 25\tfrac{1}{2} - 2\tfrac{2}{5} = 22\tfrac{2}{3} \text{ square units.}$$

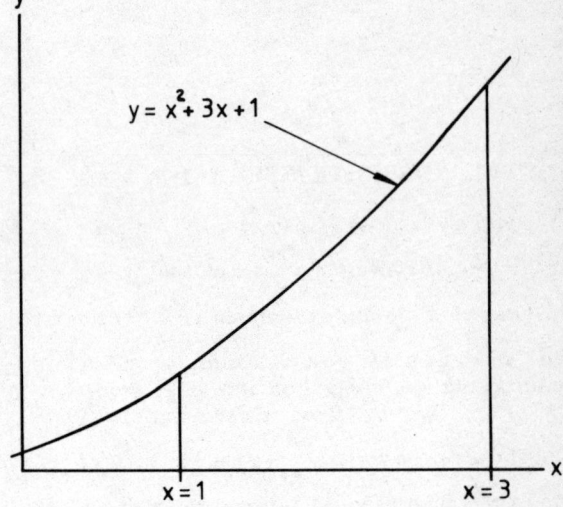

Figure 21.2

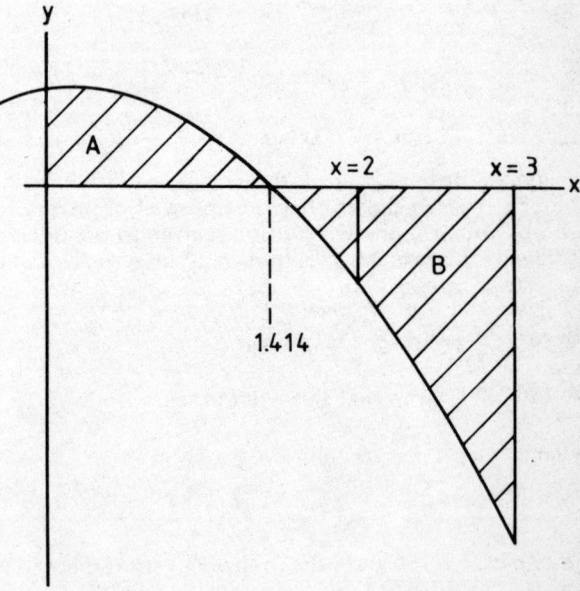

Figure 21.3

Example 21.3 Find the area between the curve $y = 2 - x^2$, the x-axis and the ordinates $x = 2$ and $x = 3$. See fig. 21.3

$$\text{Area} = \int_2^3 (2 - x^2)\,dx = \left[2x - \frac{x^3}{3}\right]_2^3$$

$$= \left(2 \times 3 - \frac{3^3}{3}\right) - \left(2 \times 2 - \frac{2^3}{3}\right)$$

$$= -4\frac{1}{3} \text{ square units.}$$

The area is found to have a negative sign, this is because it lies below the *x*-axis.

This is a general result, an area above the *x*-axis is positive and one below is negative.

In fig. 21.3 the area *A* is above the *x*-axis and the area *B* is below the *x*-axis. The curve $y = 2 - x^2$ cuts the *x*-axis at the point where $x = 1.414$.

$$\text{Area } A = \int_0^{1.414} (2 - x^2)\, dx = \left[2x - \frac{x^3}{3}\right]_0^{1.414}$$

$$= \left(2 \times 1.414 - \frac{1.414^3}{3}\right) - (0 - 0) = 1.886$$

$$\text{Area } B = \int_{1.414}^{3} (2 - x^2)\, dx = \left[2x - \frac{x^3}{3}\right]_{1.414}^{3}$$

$$= \left(2 \times 3 - \frac{3^3}{3}\right) - \left(2 \times 1.414 - \frac{1.414^3}{3}\right)$$

$$= -4.866$$

Hence total area $A + B = 1.886 + 4.886 = 6.772$

If we find $\int_0^3 (2 - x^2)\, dx = \left[2x - \frac{x^3}{3}\right]_0^3$

$$= \left(2 \times 3 - \frac{3^3}{3}\right) - (0) = -3$$

which is the difference between the areas *A* and *B*. Therefore when a curve cuts the *x*-axis we must be careful that we evaluate the area we require.

EXERCISE 21.2

1. Evaluate the following integrals:

(a) $\int_1^2 \frac{x^3 - 3}{x^2}\, dx$ (b) $\int_1^2 x^3 (x - 2)\, dx$

(c) $\int_1^3 \left(\frac{1}{x^2} + 2x + 3\right) dx$ (d) $\int_0^4 \left(\sqrt{x} - \frac{2}{\sqrt{x}}\right) dx$

(e) $\int_1^3 (x + 3)(x + 1)\, dx$ (f) $\int_0^1 (x^2 - 2)(1 - 2x^2)\, dx$

2. Find the area between the curve $y = x^2 - 2x + 5$, the *x*-axis and the ordinates $x = 1$ and $x = 3$.

3. Find the area between the curve $y = 10x - 5x^2$, the *x*-axis and the ordinates $x = 0$ and $x = 2$.

4. Find the area between the curve $y = 3 - 2x - x^2$ and the ordinates $x = 2$ and $x = 4$.

5. (a) Integrate with respect to *x*: $2x^2 - 3x + 4$

(b) Evaluate $\int_1^3 \frac{(x + 2)(x - 3) + 6}{x}\, dx$

(c) If the velocity *v* metres per second of a moving body is given in terms of the time *t* seconds by the formula $v = 2 -$

$t + \frac{1}{2}t^2$ find the distance moved during the third second.

21.3 GENERAL EXAMPLES

Example 21.4 (a) Evaluate the integrals

(i) $\int_0^2 (x - 1)(x + 5)\, dx$, (ii) $\int_1^2 \frac{x^3 - 6}{x^2}\, dx$

(b) Find *y* if $\frac{dy}{dx} = x^2 + 3$ and $y = 20$ when $x = 3$

(c) Find the area between the *x*-axis, the *y*-axis and the curve $y = 8 - x^3$.

(a) (i) $\int_0^2 (x - 1)(x + 5)\, dx = \int_0^2 (x^2 + 4x - 5)\, dx$

$$= \left[\frac{x^3}{3} + 2x^2 - 5x\right]_0^2 = \left(\frac{8}{3} + 8 - 10\right) - (0) = \frac{2}{3}$$

(ii) $\int_1^2 \frac{x^3 - 6}{x^2}\, dx = \int_1^2 (x - 6x^{-2})\, dx = \left[\frac{x^2}{2} + 6x^{-1}\right]_1^2$

$$= (2 + 3) - (\tfrac{1}{2} + 6) = -1\tfrac{1}{2}$$

(b) $\frac{dy}{dx} = x^2 + 3$ $y = \int (x^2 + 3)\, dx = \frac{x^3}{3} + 3x + c$

When $x = 3, y = 20$ hence $20 = \frac{3^3}{3} + 3 \times 3 + c$ or $c = 2$

$$y = \frac{x^3}{3} + 3x + 2$$

(c) The curve cuts the *x*-axis where $y = 0$, that is $8 - x^3 = 0$ or $x^3 = 8$, $x = 2$

$$\text{Required area} = \int_0^2 (8 - x^3)\, dx$$

$$= \left[8x - \frac{x^4}{4}\right]_0^2 = (16 - 4) - (0)$$

$$= 12 \text{ square units.}$$

EXERCISE 21.3

1. Find *y* if $\frac{dy}{dx} = \frac{2}{x^2} - 3x + 5$ and $y = 1$ when $x = 1$.

2. Integrate the following: (a) $\int \left(2 + 3\sqrt{x} - \frac{1}{x^2}\right) dx$,

(b) $\int_1^8 x^{2/3}\, dx$, (c) $\int_0^a (a^3 - x^3)\, dx$

3. (a) Integrate the following with respect to *x*:

(i) $3x^2 - 5x + 2$, (ii) $\frac{5}{x^2} + 3\sqrt{x}$

(b) Sketch the curve $y = 5x - x^2$ and show that the curve crosses the *x*-axis at $x = 0$ and $x = 5$. Calculate the area between the curve and the *x*-axis.

4. (a) Differentiate $y = \frac{5x^3}{2} - \frac{3x}{4} - \frac{2}{3} + \frac{3}{5x} + \frac{4}{3x^3}$

(b) Evaluate (i) $\int_2^3 (7x^2 + 4x - 6)\, dx$, (ii) $\int_0^1 \frac{8x^3}{5}\, dx$

(c) Find *v* if $\frac{dy}{dx} = \frac{3}{x^2} - x + 2$ and $y = 3$ when $x = 3$.

SUMMARY

1. $\int x^n dx = \dfrac{x^{n+1}}{n+1} + c$ unless $n = -1$ then

$$\int \frac{1}{x} dx = \ln x + c$$

2. $\int \sin x \, dx = -\cos x + c$, $\int \cos x \, dx = \sin x + c$

3. The area between the curve $y = f(x)$, the x-axis and the ordinates at $x = a$ and $x = b$ is given by $\int_a^b f(x) dx$

4. If the area is above the x-axis the value of the integral is positive, if the area is below the x-axis the area is negative.

SELF ASSESSMENT PAPER No 21

Instructions: Answer all questions in both sections
Time allowed: Section A 30 minutes (30 marks)
          Section B 50 minutes (25 marks each question)
Marks gained: 40+ pass with credit, 32–40 pass, less than 32 fail, repeat Chapter 21.

Section A

1. Find (a) $\int (3x^2 + 4) dx$, (b) $\int -2 \cos x \, dx$

2. Evaluate (a) $\int_1^2 (1 - x) dx$, (b) $\int_0^4 \sqrt{x} \, dx$

3. Find y if $\dfrac{dy}{dx} = 3x^2 - 7 + \dfrac{1}{x^2}$ and $y = 9$ when $x = 1$

4. Find the area between the curve $y = x^2 + 3$, the x-axis and the ordinates $x = 1$ and $x = 2$.

Section B

1. (a) Evaluate (i) $\int_1^3 (x^2 - x - 2) dx$,

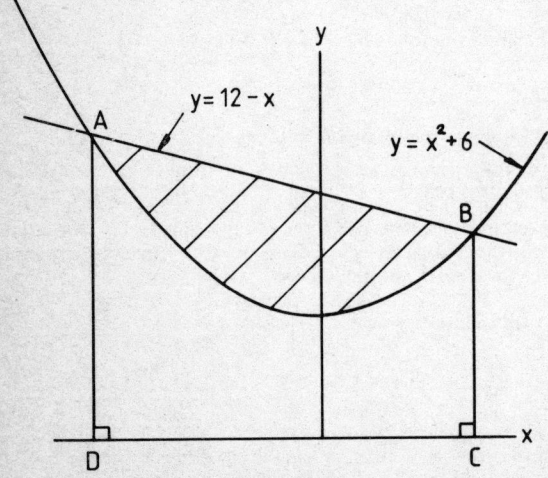

Figure 21.4

(ii) $\int_0^2 (x - 1)(x + 2) dx$

(b) The velocity of a body is given by $\dfrac{ds}{dt} = t^2 + t + 1$. Find the distance moved from the time $t = 1$ s to $t = 3$ s.

2. Fig. 21.4 shows sketches of the graphs $y = x^2 + 6$ and $y = 12 - x$. (a) By solving the simultaneous equations $y = x^2 + 6$ and $y = 12 - x$ find the co-ordinates of the points A and B where the two curves intersect.

(b) By integration find the area bounded by the curve $y = x^2 + 6$, the ordinates AD and BC and the x-axis.

(d) Hence find the area between the curve $y = x^2 + 6$ and the line $y = 12 - x$.

ANSWERS

Exercise 21.1   1. $\dfrac{7x^4}{4} - \dfrac{2x^3}{3} + 5x + c$,

2. $\dfrac{x^3}{3} - x^2 + 4x + c$,   3. $\dfrac{x^3}{3} + \dfrac{1}{x} + c$,

4. $2x - \dfrac{x^3}{15} - 2\sqrt{x} + c$,   5. $4\sqrt{x} - \dfrac{x^2}{2} + c$,

6. $\dfrac{x^{2.2}}{2.2} + \dfrac{x^{3.4}}{3.4} + c$,   7. $\dfrac{2x^{5/2}}{5} - \dfrac{7x^3}{3} + c$,

8. $3x + \dfrac{4}{x} + c$,   9. $x^3 + \dfrac{x^2}{2} - 2x + c$,

10. $\dfrac{x^4}{4} - 6x - \dfrac{4}{x^2} + c$,

11. $-2 \cos x + 3 \sin x + c$,

12. $\sin x + \cos x + c$,

Exercise 21.2   1. (a) 0, (b) -1.3, (c) $14\frac{2}{3}$, (d) $-2\frac{2}{3}$, (e) $37\frac{1}{3}$, (f) $-3\frac{2}{5}$,   2. $10\frac{2}{3}$,   3. $6\frac{2}{3}$,

4. 22,   5. (a) $\dfrac{2x^3}{3} - \dfrac{3x^2}{2} + 4x$, (b) 2, (c) $2\frac{2}{3}$

Exercise 21.3   1. $y = -\dfrac{2}{x} - \dfrac{3x^2}{2} + 5x - \dfrac{5}{2}$,

2. (a) $2x + 2\sqrt{x^3} + \dfrac{1}{x}$, (b) $18\frac{3}{5}$, (c) $\dfrac{3a^4}{4}$

3. (a) (i) $x^3 - \dfrac{5x^2}{2} + 2x$, (ii) $-\dfrac{5}{x} + 2\sqrt{x^3}$, (b) $20\frac{5}{6}$,

4. (a) $\dfrac{15x^2}{2} - \dfrac{3}{4} - \dfrac{3}{5x^2} + \dfrac{4}{x^4}$, (b) (i) $48\frac{1}{3}$, (ii) $\frac{2}{3}$,

(c) $y = -\dfrac{3}{x} - \dfrac{x^2}{2} + 2x + \dfrac{5}{2}$.

SELF ASSESSMENT PAPER No 21

Marks

Section A

1. (a) $x^3 + 4x + c$,   3
   (b) $-2 \sin x + c$,   2

2. (a) $\left[x - \dfrac{x^2}{2}\right]_1^2$ 2

$= (2 - 2) - (1 - \frac{1}{2})$ 2

$= -\frac{1}{2}$ 1

(b) $\int_0^4 x^{1/2}\,dx$ 1

$= [\frac{2}{3}x^{3/2}]_0^4$ 2

$= (\frac{16}{3}) - (0)$ 2

$= 5\frac{1}{3}$ 1

3. $y = x^3 - 7x - \dfrac{1}{x} + c$ 2

$y = 1 - 7 - 1 + c$ 2

$c = 16$ 1

$y = x^3 - 7x - \dfrac{1}{x} + 16$ 1

4. Area $= \int_1^2 (x^2 + 3)\,dx$ 2

$= \left[\dfrac{x^3}{3} + 3x\right]_1^2$ 2

$= (\frac{8}{3} + 6) - (\frac{1}{3} + 3)$ 2

$= 5\frac{1}{3}$ 2

Section B

1. (a) (i) $\left[\dfrac{x^3}{3} - \dfrac{x^2}{2} - 2x\right]_1^3$ 3

$= (9 - 4\frac{1}{2} - 6) - (\frac{1}{3} - \frac{1}{2} - 2)$ 3

$= \frac{2}{3}$ 2

(ii) $\int_0^2 (x^2 + x - 2)\,dx$ 1

$= \left[\dfrac{x^3}{3} + \dfrac{x^2}{2} - 2x\right]_0^2$ 3

$= (\frac{8}{3} + 2 - 4) - (0)$ 2

$= \frac{2}{3}$ 1

(b) $\int_1^3 (t^2 + t + 1)\,dt$ 3

$= \left[\dfrac{t^3}{3} + \dfrac{t^2}{2} + t\right]_1^3$ 3

$= (9 + 4\frac{1}{2} + 3) - (\frac{1}{3} + \frac{1}{2} + 1)$ 2

$= 14\frac{2}{3}$ 2

2. (a) $x^2 + 6 = 12 - x$ 2

$x^2 + x - 6 = 0$ 1

$(x + 3)(x - 2) = 0$ 2

$x = -3, x = 2$ 2

A $(-3, 15)$, B $(2, 10)$ 2

(b) Area $= \left(\dfrac{15 + 10}{2}\right) \times (3 + 2)$ 4

Area $= 62\frac{1}{2}$ 1

(c) Area $= \int_{-3}^2 (x^2 + 6)\,dx$ 3

$= \left[\dfrac{x^3}{3} + 6x\right]_{-3}^2$ 2

$= (\frac{8}{3} + 12) - (-9 - 18)$ 2

$= 41\frac{2}{3}$ 1

(d) Area $= 62\frac{1}{2} - 41\frac{2}{3}$ 2

$= 20\frac{5}{6}$ 1

PHASE TEST 7

Instructions: Answer all questions
Time allowed: One hour
Marks gained: 40+ pass with credit, 32–40 pass, less than 32 fail, repeat chapters 19, 20 and 21.

1. Differentiate from first principles $y = x^2 - 3x$.

2. Differentiate (a) $(x^2 - 3)(x + 2)$, (b) $x^2 - \dfrac{1}{x} + \sqrt{x}$, (c) $\sin x - \cos x$, (d) e^{-3x}.

3. Integrate (a) $7x^2 - 6x + 5$, (b) $\sqrt{x} - \dfrac{1}{x^2}$, (c) $\frac{1}{2}\cos x$, (d) e^{2x}.

4. Find the value of $\int_{-1}^2 (x - 1)(2 + x)\,dx$

5. Find the area between the curve $y = x^2 - 5x + 4$ the x-axis and the ordinates $x = 2$ and $x = 3$.

6. The current I mA flowing in a capacitor which is being discharged varies with the time as given below.

I (mA)	200	23	2.8	0.6
t (ms)	100	200	300	400

Show that these results are related by the law $I = I_0 e^{kt}$ where I_0 and k are constants. Determine values for I_0 and k.

ANSWERS PHASE TEST 7

Marks

1. $y + \delta y = (x + \delta x)^2 - 3(x + \delta x)$ 2

$\delta y = 2x\delta x + (\delta x)^2 - 3\delta x$ 2

$\dfrac{\delta y}{\delta x} = 2x + \delta x - 3$ 2

$\dfrac{dy}{dx} = 2x - 3$ 2

2. (a) $\dfrac{d}{dx}(x^3 + 2x^2 - 3x - 6)$ 3

$= 3x^2 + 4x - 3$ 3

(b) $\dfrac{d}{dx}(x^2 - x^{-1} + x^{1/2})$ 3

$= 2x + x^{-2} + \frac{1}{2}x^{-1/2}$ 2

$$= 2x + \frac{1}{x^2} + \frac{1}{2\sqrt{x}}$$ 2

(c) $\frac{d}{dx}(\sin x - \cos x) = \cos x + \sin x$ 2

(d) $\frac{d}{dx}(e^{-3x}) = -3e^{-3x}$ 2

3. (a) $\int (7x^2 - 6x + 5)\,dx = \frac{7x^3}{3} - \frac{6x^2}{2} + 5x + c$ 4

$$= \frac{7}{3}x^3 - 3x^2 + 5x + c$$ 1

(b) $\int (x^{1/2} - x^{-2})\,dx = \frac{x^{3/2}}{\frac{3}{2}} - \frac{x^{-1}}{-1} + c$ 3

$$= \frac{2}{3}\sqrt{x^3} + \frac{1}{x} + c$$ 2

(c) $\int \frac{1}{2}\cos x\,dx = \frac{1}{2}\sin x + c$ 2

(d) $\int e^{2x}\,dx = \frac{1}{2}e^{2x} + c$ 2

4. $\int_{-1}^{2}(x^2 + x - 2)\,dx$ 3

$$= \left[\frac{x^3}{3} + \frac{x^2}{2} - 2x\right]_{-1}^{2}$$ 3

$$= (\tfrac{8}{3} + \tfrac{4}{2} - 4) - (-\tfrac{1}{3} + \tfrac{1}{2} + 2)$$ 2

$$= \tfrac{2}{3} - (2\tfrac{1}{6}) = -1\tfrac{1}{2}$$ 2

5. Area $= \int_{2}^{3}(x^2 - 5x + 4)\,dx$ 2

$$= \left[\frac{x^3}{3} - \frac{5x^2}{2} + 4x\right]_{2}^{3}$$ 3

$$= (9 - \tfrac{45}{2} + 12) - (\tfrac{8}{3} - 10 + 8)$$ 2

$$= -1\tfrac{1}{2} - \tfrac{2}{3} = -2\tfrac{1}{6}$$ 2

Area $= 2\tfrac{1}{6}$ square units   1

6. $\ln I = kt + \ln I_0$ 2

t	100	200	300	400	
$\ln I$	5.30	3.14	1.03	-0.51	4

Graph of $\ln I$ against t 6

Intercept on the $\ln I$ axis $\ln I_0 = 7.1$ 3

$$I_0 = 1210$$ 2

Gradient $k = \dfrac{-7.1}{370} = -0.019$ 3

22 Vectors

22.1 VECTORS AT A POINT

A vector is a physical quantity which has both magnitude and direction.

Thus speed is a scalar quantity but velocity is a vector quantity;

    distance is a scalar quantity but displacement is a vector quantity;

    force is also a vector quantity.

Example 22.1 A body is being pulled by a force of 80N making an angle of 60˘ to the horizontal, as shown in fig. 22.1.

Find the horizontal and vertical components of the force. The force can be divided into two parts a horizontal part and a vertical part. The horizontal part will be the projection of OP onto the horizontal axis. This will be 80 cos 60°. The vertical part will be the projection of OP onto the vertical axis. This will be 80 cos ∠BOP. But cos ∠BOP = sin ∠POA because ∠BOA = 90°. Hence the vertical part will be 80 sin 60°.

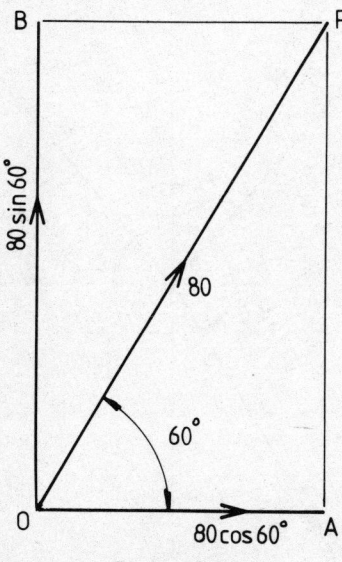

Figure 22.1

The vertical component of the force is 80 sin 60°.
The horizontal component of the force is 80 cos 60°.

In general any vector can be divided into two parts at right angles. If the angle between the vector F and the horizontal axis is θ then F cos θ is called the resolved part of the vector in the horizontal direction and F sin θ is called the resolved part of the vector in the vertical direction.

If several forces are acting at a point then the resultant of these forces can be found by adding the resolved parts of the forces in the two directions.

Example 22.2 Two forces 100 N at 30° to the horizontal and 60 N at 70° to the horizontal are acting at the point O. Find the resultant of these two forces.

In fig. 22.2, the vector $\overrightarrow{OP}$ has components OA = 100 cos 30° and OB = 70 sin 30˘ and the vector $\overrightarrow{OQ}$ has components OC = 60 cos 70° and OD = 60 sin 70°.

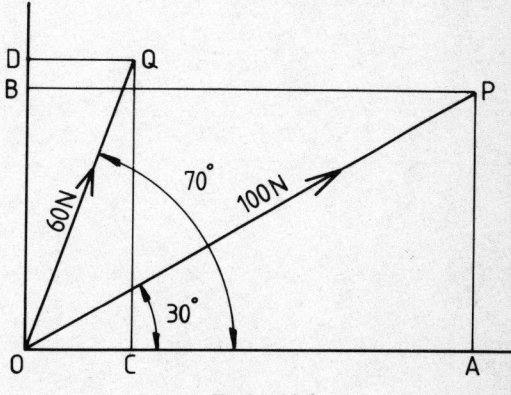

Figure 22.2

Now the components can be added:

The total horizontal component = OA + OC = 100 cos 30° + 60 cos 70° = 86.60 + 20.52 = 107.12 N

The total vertical component = OB + OD = 100 sin 30° + 60 sin 70° = 50.00 + 55.38 = 106.38 N

The resultant of these two components can now be found. If the resultant is OR, see fig. 22.3, then $OR^2 = OE^2 + OF^2 = 107.12^2 + 106.38^2 = 22\,791$

$$OR = \sqrt{22\,791} = 151\,N$$

$$\tan \theta = \frac{ER}{OE} = \frac{OF}{OE} = \frac{106.38}{107.12} = 0.9931$$

$$\theta = 44.80°$$

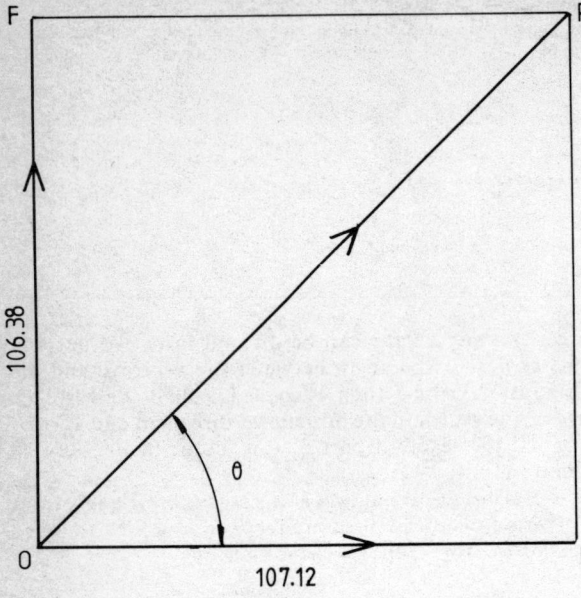

Figure 22.3

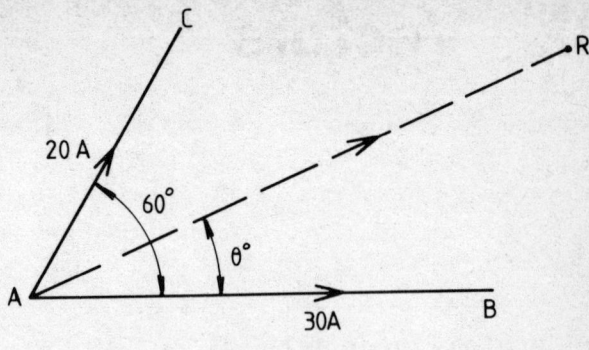

Figure 22.5

The resultant of the two forces is a force 151 N at 44.80° to the horizontal. In general any vector quantity is made up of two parts, a length and an angle. Thus in fig. 22.4 the vector $\overrightarrow{OP}$ has a length OP and makes an angle of θ with the horizontal OA.

$$OP = \sqrt{x^2 + y^2}, \ \tan \theta = \frac{y}{x}$$

In this example we can resolve the vectors in directions AB and at right angles to AB.

Resolving along AB we have AB + AC cos 60°

$$= 30 + 20 \cos 60° = 40A$$

At right angles to AB we have AB sin 90° + AC sin 60°

$$= 0 + 20 \sin 60° = 17.32 \ A$$

If the resultant is a vector AR then

$$AR^2 = 40^2 + 17.32^2 = 1900, \ AR = \sqrt{1900} = 43.6 \ A$$
$$\tan \theta = \frac{17.32}{40} = 0.433, \ \theta = 23.4°$$

The resultant current is 43.6 A at an angle of 23.4° with AB.

In the examples we have looked at all of the angles

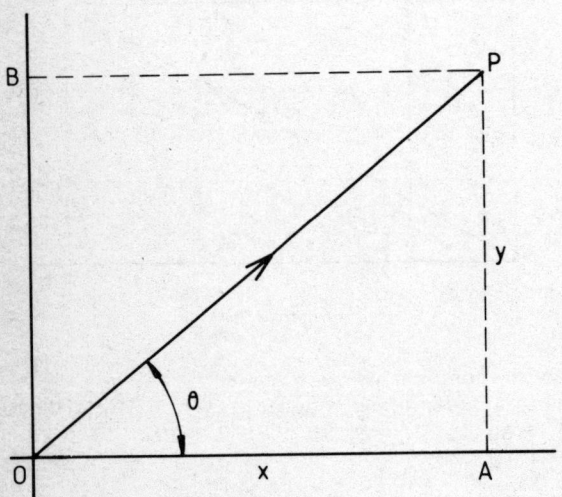

Figure 22.4

Example 22.3 In fig. 22.5 the vectors AB and AC represent the alternating currents in two branches of a parallel circuit. Calculate the resultant current and the angle it makes with AB.

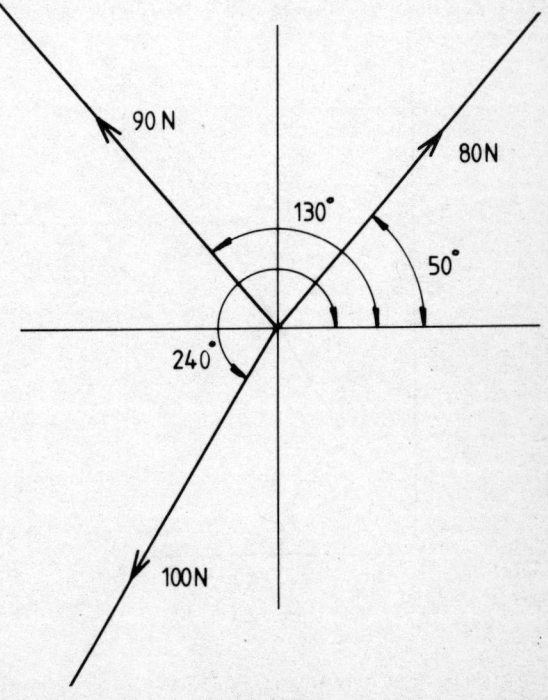

Figure 22.6

have been between $0°$ and $90°$. The same method applies to any vector $\overrightarrow{OP}$ at angle θ to horizontal, the horizontal component is OP cos θ and the vertical component is OP sin θ.

Example 22.4 Three forces act at the point O as shown in fig. 22.6. Find the resultant of these forces.
Resolving horizontally

$80 \cos 50° + 90 \cos 130° + 100 \cos 240°$

$= 80 \cos 50° - 90 \cos 50° - 100 \cos 60°$

$= 51.42 - 57.85 - 50 = -56.43$

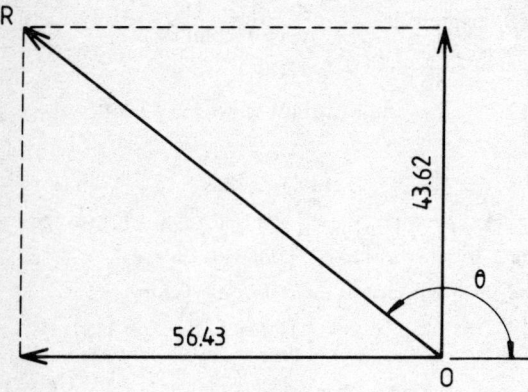

Figure 22.7

Resolving vertically

$80 \sin 50° + 90 \sin 130° + 100 \sin 240°$

$= 80 \sin 50° + 90 \sin 50° - 100 \sin 60°$

$= 61.28 + 68.94 - 86.60 = 43.62$

The resultant of these is OR, shown in fig. 22.7.

$OR^2 = 56.43^2 + 43.62^2 + 5087$

$OR = \sqrt{5087} = 71.3 \, N$

$\tan (180° - \theta°) = \dfrac{43.62}{56.43} = 0.7730$

$180° - \theta° = 37.70° \quad \theta = 180° - 37.70° = 142.30°$

The resultant force is 71.3 N at an angle of $142.30°$.

EXERCISE 22.1

1. Find the horizontal and vertical components of the following forces (a) 20 N at an angle of $60°$ to the horizontal (b) 90 N at an angle of $70°$ to the horizontal, (c) 50 N at an angle of $20°$ to the horizontal.

2. If two forces of 120 N at an angle of $30°$ to the horizontal and 80 N at an angle of $60°$ to the horizontal are acting at the point O, find the resultant force.

3. Find the resultant of the two vectors $\overrightarrow{OA}$, $\overrightarrow{OB}$ if OA = 30, OB = 20 and angle AOB = $40°$.

4. Find the resultant of the forces 60 N at $20°$ to the horizontal, 70 N at $80°$ to the horizontal and 50 N at $50°$ to the horizontal.

5. The vectors $\overrightarrow{AB}$ and $\overrightarrow{AC}$ represent the alternating currents in two branches of a parallel circuit. If AB = 20 A, AC = 15 A and $\angle BAC = 60°$ calculate the resultant current and the angle it makes with AB.

6. Three forces act at the point O. If the magnitude of the force and the direction it makes with the horizontal in each case is 50 N at $40°$, 60 N at $120°$ and 70 N at $250°$ find the resultant force.

22.2 ADDITION AND SUBTRACTION OF VECTORS

In the previous paragraph we looked at the addition of vectors by adding and subtracting the resolved parts of a vector. We now look at the more general problem. A vector can be represented by its magnitude and its direction or it can be represented by the resolved parts of the vector. Thus if we refer back to fig. 22.1 we can represent the vector OP by 80 N at $60°$ or $(80 \cos 60°, 80 \sin 60°)$ which are the x and y components if we refer to the normal x and y axes.

If we are using the x and y components to represent the vector it is represented by the matrix $\begin{pmatrix} x \\ y \end{pmatrix}$. Thus the vector $\overrightarrow{OP}$ in fig. 22.1 is the vector $\begin{pmatrix} 80 \cos 60° \\ 80 \sin 60° \end{pmatrix}$.

Now if we have two vectors $\begin{pmatrix} 2 \\ 1 \end{pmatrix}$ and $\begin{pmatrix} 2 \\ 4 \end{pmatrix}$ we have by matrix addition that $\begin{pmatrix} 2 \\ 1 \end{pmatrix} + \begin{pmatrix} 2 \\ 4 \end{pmatrix} = \begin{pmatrix} 4 \\ 5 \end{pmatrix}$

Hence in fig. 22.8 we see that the addition of these vectors is the diagonal of the parallelogram formed by the vectors.

$$\overrightarrow{OA} = \begin{pmatrix} 2 \\ 1 \end{pmatrix}, \overrightarrow{AB} = \begin{pmatrix} 2 \\ 4 \end{pmatrix}, \overrightarrow{OB} = \begin{pmatrix} 4 \\ 5 \end{pmatrix}$$

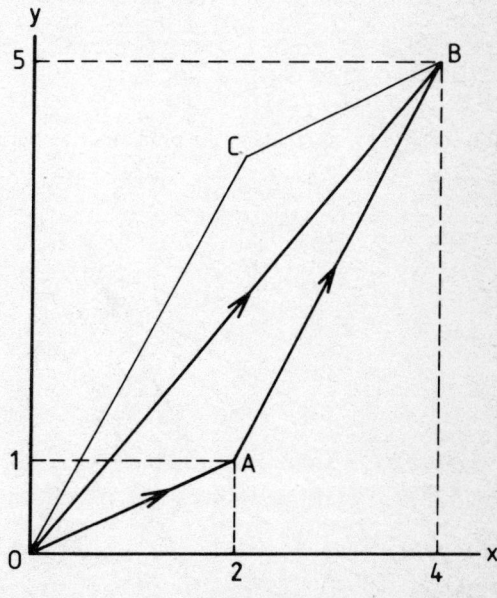

Figure 22.8

From fig. 22.8 if, in general, $\overrightarrow{OA} = \mathbf{a}$, and $\overrightarrow{AB} = \mathbf{b}$ then $\overrightarrow{OC} = \mathbf{b}, \overrightarrow{CB} = \mathbf{a}, \overrightarrow{OB} = \mathbf{a} + \mathbf{b}$. Also $\overrightarrow{AO} = -\mathbf{a}, \overrightarrow{CO} = -\mathbf{b}$

and hence $\overrightarrow{AC} = \overrightarrow{AO} + \overrightarrow{OC} = -a + b = b - a$ and $\overrightarrow{CA} = \overrightarrow{CO} + \overrightarrow{OA} = -b + a = a - b$.

In practice the vectors are seldom simple whole numbers and hence the diagonal of the parallelogram must be found by using the sine or cosine rule.

Example 22.5 If $a = \begin{pmatrix} 3 \\ 2 \end{pmatrix}$, $b = \begin{pmatrix} 2 \\ 3 \end{pmatrix}$ and $c = \begin{pmatrix} -1 \\ 2 \end{pmatrix}$, find

(a) (i) $a + b$, (ii) $\frac{1}{2}(a - c)$, (iii) $2a - 3b + c$.

(b) the magnitude of a and c

(c) k and n such that $ka + nb = c$

(a) (i) $a + b = \begin{pmatrix} 3 \\ 2 \end{pmatrix} + \begin{pmatrix} 2 \\ 3 \end{pmatrix} = \begin{pmatrix} 5 \\ 5 \end{pmatrix}$

(ii) $\frac{1}{2}(\begin{pmatrix} 3 \\ 2 \end{pmatrix} - \begin{pmatrix} -1 \\ 2 \end{pmatrix}) = \frac{1}{2}\begin{pmatrix} 4 \\ 0 \end{pmatrix} = \begin{pmatrix} 2 \\ 0 \end{pmatrix}$

(iii) $2a - 3b + c = 2\begin{pmatrix} 3 \\ 2 \end{pmatrix} - 3\begin{pmatrix} 2 \\ 3 \end{pmatrix} + \begin{pmatrix} -1 \\ 2 \end{pmatrix} = \begin{pmatrix} -1 \\ -3 \end{pmatrix}$

(b) Magnitude of a is $\sqrt{3^2 + 2^2} = \sqrt{13} = 3.606$

Magnitude of c is $\sqrt{(-1)^2 + 2^2} = \sqrt{5} = 2.236$

(c) $k\begin{pmatrix} 3 \\ 2 \end{pmatrix} + n\begin{pmatrix} 2 \\ 3 \end{pmatrix} + \begin{pmatrix} -1 \\ 2 \end{pmatrix}$

$\begin{pmatrix} 3k + 2n \\ 2k + 3n \end{pmatrix} = \begin{pmatrix} -1 \\ 2 \end{pmatrix}$ $\quad$ $3k + 2n = -1$ (1)
$\quad\quad\quad\quad\quad\quad\quad\quad\quad\quad\quad 2k + 3n = 2$ (2)

To solve $2 \times (1) - 3 \times (2)$

$\quad\quad 4n - 9n = -2 - 6$
$\quad\quad\quad -5n = -8$
$\quad\quad\quad\quad\quad n = \dfrac{8}{5}$

substitute $n = \dfrac{8}{5}$ into (1) $\quad 3k + \dfrac{16}{5} = -1$

$\quad\quad\quad\quad\quad\quad\quad\quad\quad\quad\quad k = -\dfrac{7}{5}$

$n = \dfrac{8}{5}, k = -\dfrac{7}{5}$

Example 22.6 Two currents in the branches of a parallel

alternating current circuit are represented by the vectors $22\angle 30°$ and $16\angle 60°$. Find the resultant as a vector. In fig. 22.9 $\overrightarrow{OA}$ represents the vector $22\angle 30°$ and $\overrightarrow{OB}$ the vector $16\angle 60°$.

The resultant is represented by $\overrightarrow{OC}$. In $\triangle OAC$, $\angle OAC = 150°$, $OA = 22$, $AC = 16$.

Using the cosine rule

$OC^2 = 22^2 + 16^2 - 2 \times 22 \times 16 \times \cos 150°$

$OC^2 = 484 + 256 + 610 = 1350$

$OC = \sqrt{1350} = 36.7$

Using the sine rule $\quad \dfrac{36.7}{\sin 150} = \dfrac{16}{\sin \theta}$

$\sin \theta = \dfrac{16 \times \sin 150}{36.7} = 0.2180$

$\theta = 12.59°$ $\quad$ The resultant is $36.7\angle 42.59°$

EXERCISE 22.2

1. If $a = \begin{pmatrix} 3 \\ 2 \end{pmatrix}$, $b = \begin{pmatrix} 1 \\ -2 \end{pmatrix}$, $c = \begin{pmatrix} -2 \\ 0 \end{pmatrix}$ find (a) $a - b$, (b) $2b - c$, (c) $\frac{1}{2}(a - b)$, (d) $a + b - c$, (e) $3a - 2b + c$

2. Find the magnitude of the following vectors
(a) $\begin{pmatrix} 1 \\ 1 \end{pmatrix}$, (b) $\begin{pmatrix} 2 \\ 3 \end{pmatrix}$, (c) $\begin{pmatrix} -2 \\ 1 \end{pmatrix}$, (d) $\begin{pmatrix} -2 \\ -3 \end{pmatrix}$, (e) $\begin{pmatrix} 0 \\ 2 \end{pmatrix}$

3. If $a + kb = c$ find k when (a) $a = \begin{pmatrix} 2 \\ 1 \end{pmatrix}$, $b = \begin{pmatrix} -1 \\ 2 \end{pmatrix}$, $c = \begin{pmatrix} -1 \\ 7 \end{pmatrix}$, (b) $a = \begin{pmatrix} 2 \\ 2 \end{pmatrix}$, $b = \begin{pmatrix} -2 \\ 0 \end{pmatrix}$, $c = \begin{pmatrix} 1 \\ 2 \end{pmatrix}$

4. If $ka + nb = c$ find k and n if $a = \begin{pmatrix} 2 \\ 0 \end{pmatrix}$, $b = \begin{pmatrix} 3 \\ 2 \end{pmatrix}$ and $c = \begin{pmatrix} 5 \\ 4 \end{pmatrix}$.

5. Two forces are represented by the vectors $90\angle 20°$ and $80\angle 70°$ find the resultant force as a vector.

6. Two currents in the branches of a parallel alternating current circuit are represented by the vectors $20\angle 15°$ and $40\angle 75°$. Find the resultant vector.

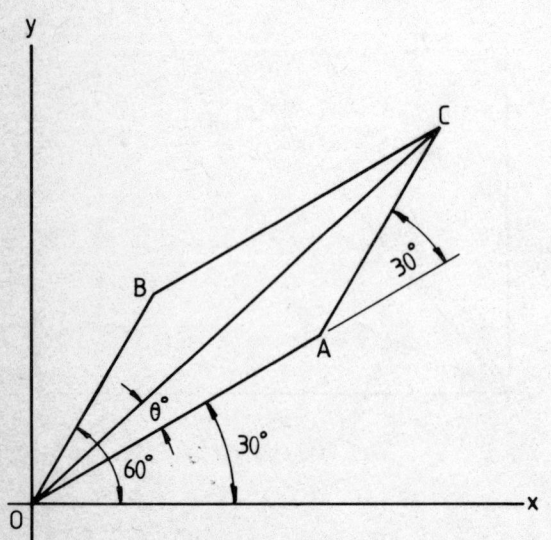

Figure 22.9

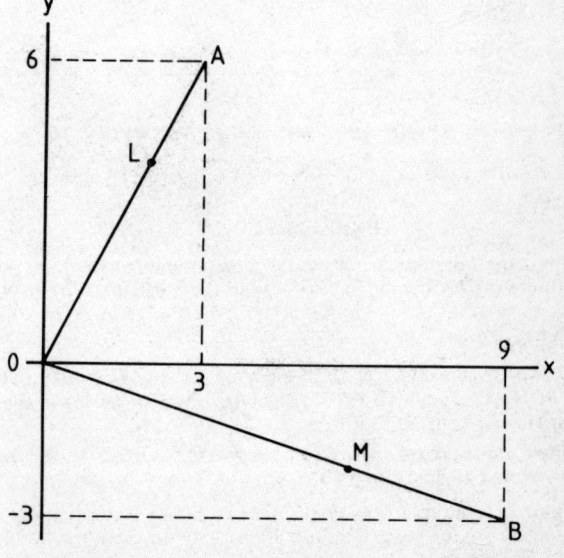

Figure 22.10

22.3 GENERAL EXAMPLES

Example 22.7 The position vectors of A and B relative to O are $\begin{pmatrix}3\\6\end{pmatrix}$ and $\begin{pmatrix}9\\-3\end{pmatrix}$ respectively. Points L, M divide OA OB in the ratio 2:1. Find (i) the position vectors of L and M, (ii) $\overrightarrow{AB}$, (iii) $\overrightarrow{LM}$. Show that $\overrightarrow{AB}$ is parallel to $\overrightarrow{LM}$ and find the ratio LM:AB

The points A, B, L, M are as shown in fig. 22.10

$OL = \frac{2}{3}$ of OA hence it is obvious that $\overrightarrow{OL} = \frac{2}{3}\overrightarrow{OA}$ similarly $\overrightarrow{OM} = \frac{2}{3}\overrightarrow{OB}$

(i) $\overrightarrow{OL} = \frac{2}{3}\begin{pmatrix}3\\6\end{pmatrix} = \begin{pmatrix}2\\4\end{pmatrix}$, $\overrightarrow{OM} = \frac{2}{3}\begin{pmatrix}9\\-3\end{pmatrix} = \begin{pmatrix}6\\-2\end{pmatrix}$

(ii) $\overrightarrow{AB} = \overrightarrow{AO} + \overrightarrow{OB} = -\overrightarrow{OA} + \overrightarrow{OB} = -\begin{pmatrix}3\\6\end{pmatrix} + \begin{pmatrix}9\\-3\end{pmatrix}$
$= \begin{pmatrix}6\\-9\end{pmatrix}$

(iii) $\overrightarrow{LM} = \overrightarrow{LO} + \overrightarrow{OM} = -\overrightarrow{OL} + \overrightarrow{OM} = -\begin{pmatrix}2\\4\end{pmatrix} + \begin{pmatrix}6\\-2\end{pmatrix}$
$= \begin{pmatrix}4\\-6\end{pmatrix}$

(iv) $\overrightarrow{AB} = \begin{pmatrix}6\\-9\end{pmatrix} = 3\begin{pmatrix}2\\-3\end{pmatrix}$, $\overrightarrow{LM} = \begin{pmatrix}4\\-6\end{pmatrix} = 2\begin{pmatrix}2\\-3\end{pmatrix}$

If $\mathbf{x} = \begin{pmatrix}2\\-3\end{pmatrix}$ then $\overrightarrow{AB} = 3\mathbf{x}$ and $\overrightarrow{LM} = 2\mathbf{x}$

Hence $\overrightarrow{AB}$ and $\overrightarrow{LM}$ are parallel and LM:AB = 2:3

EXERCISE 22.3

1. An aircraft flies from an airport A, position vector relative to an origin O is $\begin{pmatrix}50\\25\end{pmatrix}$ km. The velocity of the aircraft is $\begin{pmatrix}12\\15\end{pmatrix}$ km/min. Find the position vector of the aircraft relative to O after 10 minutes.

2. A man in a boat wishes to travel directly across a river in which there is a current of 0.8 m/s. The boat travels at 2.0 m/s. Find the angle to the flow of the river the boat should head in order to travel directly across the river.

3. The position vectors of points A and B relative to O are $\begin{pmatrix}4\\8\end{pmatrix}$ and $\begin{pmatrix}12\\4\end{pmatrix}$ respectively. Points L and M divide OA, OB in the ratio 3:1. Find (a) the position vectors of L and M, (b) $\overrightarrow{AB}$, (c) $\overrightarrow{LM}$. Show that $\overrightarrow{AB}$ is parallel to $\overrightarrow{LM}$, and find the ratio of the lengths LM:AB.

4. Two forces are represented by the vectors $70\angle 30°$ and $90\angle 150°$. Find the resultant force as a vector.

SUMMARY

1. A vector is a physical quantity which has magnitude and direction. It is represented by $\overrightarrow{AB}$, a or $\begin{pmatrix}x\\y\end{pmatrix}$.

2. If a vector $\overrightarrow{OA}$ is of length a and makes an angle of $\theta°$ with the horizontal then the resolved part of $\overrightarrow{OA}$ horizontally is $a\cos\theta°$ and vertically is $a\sin\theta°$.

3. The resultant of several vectors can be found by adding together the resolved parts of the vectors.

4. The magnitude of the vector $\begin{pmatrix}x\\y\end{pmatrix}$ is $\sqrt{x^2 + y^2}$, the angle it makes with the x-axis is θ where $\tan\theta = \frac{y}{x}$.

5. If two vectors are represented by the sides of a

parallelogram then the resultant is represented by the diagonal of the parallelogram.

SELF ASSESSMENT PAPER No 22

Instructions: Answer all questions in both sections
Time allowed: Section A 20 minutes (20 marks)
Section B 40 minutes (20 marks each question)
Marks gained: 30+ pass with credit, 24–30 pass, less than 24 fail, repeat chapter 22.

Section A

1. A vector has length 6 and makes an angle of 60° with the x-axis. Find the resolved parts of the vector along the x and y axes.

2. If $\mathbf{a} = \begin{pmatrix}3\\-2\end{pmatrix}$ and $\mathbf{b} = \begin{pmatrix}-2\\1\end{pmatrix}$ find (a) $\mathbf{a} + \mathbf{b}$, (b) $2\mathbf{a} - \mathbf{b}$.

3. If $\mathbf{a} = \begin{pmatrix}3\\4\end{pmatrix}$ find the magnitude and direction of the vector $\mathbf{a}$.

4. Find which vectors of the following are parallel $\begin{pmatrix}1\\-1\end{pmatrix}$, $\begin{pmatrix}-1\\0\end{pmatrix}$, $\begin{pmatrix}2\\3\end{pmatrix}$, $\begin{pmatrix}0\\-1\end{pmatrix}$, $\begin{pmatrix}2\\-2\end{pmatrix}$.

5. If $\mathbf{a} = \begin{pmatrix}2\\3\end{pmatrix}$, $\mathbf{b} = \begin{pmatrix}-1\\2\end{pmatrix}$ and $\mathbf{c} = \begin{pmatrix}-1\\9\end{pmatrix}$ find a value for k such that $\mathbf{a} + k\mathbf{b} = \mathbf{c}$.

Section B

1. The position of the two points A and B relative to O are $\mathbf{a}$ and $\mathbf{b}$. If $\mathbf{a} = \begin{pmatrix}6\\-3\end{pmatrix}$ and $\mathbf{b} = \begin{pmatrix}12\\6\end{pmatrix}$ find the position vectors of 2 points L and M which divide OA and OB in the ratio 1:2. Show that $\overrightarrow{AB}$ and $\overrightarrow{LM}$ are parallel.

2. Two forces are represented by the vectors $50\angle 45°$ and $70\angle 75°$ find the resultant force as a vector.

ANSWERS

Exercise 22.1 (a) 10 N, 17.32 N. (b) 30.78 N, 84.57 N, (c) 46.98 N, 17.10 N, 2. 193, 41°56′, 3. 47, 15°50′ to OA, 4. 162.7, 51.76°, 5. 30.4 A, 25°17′ to AB, 6. 24.1 N, 130°29′.

Exercise 22.2 1. (a) $\begin{pmatrix}2\\4\end{pmatrix}$, (b) $\begin{pmatrix}4\\-4\end{pmatrix}$, (c) $\begin{pmatrix}1\\2\end{pmatrix}$, (d) $\begin{pmatrix}6\\0\end{pmatrix}$, (e) $\begin{pmatrix}5\\10\end{pmatrix}$ 2. (a) 1.414, (b) 3.606, (c) 2.236, (d) 3.606, (e) 2, 3. (a) 3, (b) ½, 4. $n = 2$, $k = -½$ 5. 154, 43°27′, 6. 52.9, 55°53′.

Exercise 22.3 1. $\begin{pmatrix}170\\175\end{pmatrix}$, 2. 66°25′, 3. (a) $\begin{pmatrix}3\\6\end{pmatrix}$, $\begin{pmatrix}9\\-3\end{pmatrix}$, (b) $\begin{pmatrix}8\\-12\end{pmatrix}$, (c) $\begin{pmatrix}6\\-9\end{pmatrix}$, 3:4, 4. 81.85, 102°13′.

SELF ASSESSMENT PAPER NO 22

Section A	Marks
1. Along x-axis $6\cos 60° = 3$	2
Along y-axis $6\sin 60° = 5.196$	2
2. (a) $\begin{pmatrix}1\\-1\end{pmatrix}$	2

(b) $\begin{pmatrix} 8 \\ -5 \end{pmatrix}$ 3

3. Magnitude $= \sqrt{3^2 + 4^2} = 5$ 3
Direction $= \tan^{-1} \frac{4}{3} = 53°7'$ 3

4. $\begin{pmatrix} 1 \\ -1 \end{pmatrix}$ and $\begin{pmatrix} 2 \\ -2 \end{pmatrix}$ are parallel   2

5. $2 - k = -1, k = 3$
check $3 + 2k = 9, k = 3$ 3

Section B

1. $\overrightarrow{OL} = \frac{1}{3}\overrightarrow{OA}$ 2

$l = \begin{pmatrix} 2 \\ -1 \end{pmatrix}$ 2

$\overrightarrow{OM} = \frac{1}{3}\overrightarrow{OB}$ 2

$m = \begin{pmatrix} 4 \\ 2 \end{pmatrix}$ 2

$\overrightarrow{AB} = \overrightarrow{AO} + \overrightarrow{OB}$ 2

$= -\begin{pmatrix} 6 \\ -3 \end{pmatrix} + \begin{pmatrix} 12 \\ 6 \end{pmatrix}$ 2

$= \begin{pmatrix} 6 \\ 9 \end{pmatrix}$ 1

$\overrightarrow{LM} = \overrightarrow{LO} + \overrightarrow{OM}$ 2

$= -\begin{pmatrix} 2 \\ -1 \end{pmatrix} + \begin{pmatrix} 4 \\ 2 \end{pmatrix}$ 2

$= \begin{pmatrix} 2 \\ 3 \end{pmatrix}$ 1

$\overrightarrow{LM} = \frac{1}{3}\overrightarrow{AB}$ 1
Hence $\overrightarrow{LM}$ is parallel to $\overrightarrow{AB}$ 1

2. Let resultant be $R\angle(\theta + 45)$, θ is angle
resultant makes with force 50   1
$R^2 = 50^2 + 70^2 - 2 \times 50 \times 70 \cos 150°$ 4
$R = 116$ 4
$\dfrac{R}{\sin 150°} = \dfrac{70}{\sin \theta}$ 3
$\sin \theta = \dfrac{70 \times \sin 150}{116}$ 3
$\theta = 17°34'$ 2
$\theta + 45 = 62°34'$ 2
Resultant is $116\angle 62°34'$ 1

EXAMINATION 3

Instructions: Answer all questions in Section A and
       4 questions in Section B
Time allowed: Section A 40 minutes (40 marks)
Section B 80 minutes (20 marks each question)
each question)
Marks gained: 60+ pass with credit, 48–60 pass, less
than 48 fail, repeat chapters 15–22

Section A

1. Factorise the following (i) $2x^2 - 6x$, (ii) $x^2 - x - 6$

2. Solve the equations (i) $2x^2 - 7x + 3 = 0$, (ii) $2x^2 - 8x + 3 = 0$

3. Find the values of (i) $\log_3 81$, (ii) $\log_5 5$

4. Find the inverse of $\begin{pmatrix} 4 & 1 \\ 5 & 3 \end{pmatrix}$

5. Differentiate (i) $\dfrac{x^2 + 3x + 5}{x}$, (ii) $e^{-x} + \cos x$

6. Integrate (i) $3x^2 - 8x + 6$, (ii) $3 \cos x - 2 \sin x$

7. Find the resolved parts of the vector $\overrightarrow{AB}$ if it is of magnitude 7 and makes an angle of $60°$ with the horizontal.

Section B

1. A bar of rectangular cross-section 75 mm by 50 mm has to be reduced to a cross-sectional area of 3036 mm² by milling a flat on each side. If the depth of cut is x millimetres show that the new area is given by $3750 - 250x + 4x^2$. Hence calculate the depth of cut, x, required.

2. The intensity of radiation, R, from a certain radioactive source at a particular time t seconds is thought to follow the law $R = kt^n$ where k and n are constants. In an experiment to test this law the following values were obtained.

R	58	43.5	26.5	14.5	10
t	1.5	2	3		

Draw a straight line graph to verify this and find the values of k and n.

3. The resistance of an element varies with temperature according to the law $R = aT + b$. On testing it was found that the resistance $R = 26.75\ \Omega$ at $T = 20°C$ and $R = 28.10\ \Omega$ at $T = 50°C$. Form a pair of simultaneous equations and solve these equations by using matrices to determine the values of a and b.

4. (a) Differentiate with respect to x: (i) $\dfrac{(x^2 - 2x)^2}{x^2}$, (ii) $3 \sin x - \cos x$, (iii) $3e^{4x}$.

(b) An open box has to have a square base of side x cm and a fixed surface area 20 cm². Find the depth, h cm, of the box if the volume is to be a maximum.

5. (a) Perform the following integrations:
(i) $\int x^2\,dx$, (ii) $\int (x^3 + 6x)\,dx$, (iii) $\int_1^4 \sqrt{x}\,dx$
(b) Calculate the area between the curve $y = x^2 + 6$, the x-axis and the ordinates at $x = 1$ and $x = 3$.

6. Two alternating voltages are given by $V_1 = 16 \sin \theta$ volts and $V_2 = 24 \sin (\theta + \frac{\pi}{6})$ volts. Determine an expression for the resultant $V_R = V_1 + V_2$.

ANSWERS EXAMINATION 3

Section A

Marks

1. (i) $2x(x-3)$ 2
   (ii) $(x+2)(x-3)$ 2

2. (i) $(2x-1)(x-3)=0$ 2
   $x=\frac{1}{2}$ or $x=3$ 2

   (ii) $x=\dfrac{8\pm\sqrt{64-24}}{4}$ 3

   $x=\dfrac{8\pm\sqrt{40}}{4}=\dfrac{8\pm6.325}{4}$ 1

   $x=3.58$ or $x=0.42$ 2

3. (i) $\log_3 3^4 = 4\log_3 3 = 4$ 2
   (ii) $\log_5 5 = 1$ 1

4. Determinant $= 4\times3 - 1\times5 = 7$ 2

   Inverse $=\dfrac{1}{7}\begin{pmatrix}3 & -1\\ -5 & 4\end{pmatrix}$ 3

5. (i) $\dfrac{d}{dx}(x+3+5x^{-1})$ 1

   $=1+0-5x^{-2}$ 2

   $=1-\dfrac{5}{x^2}$ 1

   (ii) $\dfrac{d}{dx}(e^{-x}+\cos x)=-e^{-x}-\sin x$ 2

6. (i) $\int(3x^2-8x+6)dx=\dfrac{3x^3}{3}-\dfrac{8x^2}{2}+6x+c$ 3

   $=x^3-4x^2+6x+c$ 1

   (ii) $\int(3\cos x-2\sin x)dx=3\sin x+2\cos x+c$ 2

7. Horizontal component $=7\cos 60° = 3.5$ 3
   Vertical component $=7\sin 60° = 6.062$ 3

Section B

1. New cross-sectional area $=(75-2x)(50-2x)$ 4
   $=3750-250x+4x^2$ 2
   $3750-250x+4x^2=3036$ 2
   $4x^2-250x+714=0$ 1
   $2x^2-125x+357=0$ 1

   $x=\dfrac{125\pm\sqrt{125^2-4\times2\times357}}{2\times2}$ 5

   $x=\dfrac{125\pm\sqrt{12\,769}}{4}=\dfrac{125\pm113}{4}$ 2

   $x=59.5$ or $x=3$ 2
   Hence depth of cut required is 3 mm 1

2. $\log_{10}R=n\log_{10}t+\log_{10}k$ 2

$\log_{10}R$	1.763	1.638	1.423	1.161	1.000
$\log_{10}t$	0.176	0.301	0.477	0.699	0.845

   5

Graph of $\log_{10}R$ against $\log_{10}t$ 4
Two points from the graph $\log_{10}R=1.80$,
$\log_{10}t=0.14$, 2
$\log_{10}R=1.16$, $\log_{10}t=0.70$ 2
$1.80=0.14n+\log_{10}k$ (1) 1
$0.70=1.16n+\log_{10}k$ (2) 1
Solving $n=-1.08$ 2
$\log_{10}k=1.943$, $k=87.7$ 3

3. $26.75=20a+b$ 2
   $28.10=50a+b$ 2

   $\begin{pmatrix}50 & 1\\ 20 & 1\end{pmatrix}\begin{pmatrix}a\\ b\end{pmatrix}=\begin{pmatrix}28.1\\ 26.75\end{pmatrix}$ 4

   Inverse $=\dfrac{1}{30}\begin{pmatrix}1 & -1\\ -20 & 50\end{pmatrix}$ 5

   $\begin{pmatrix}a\\ b\end{pmatrix}=\dfrac{1}{30}\begin{pmatrix}1 & -1\\ -20 & 50\end{pmatrix}\begin{pmatrix}28.1\\ 26.75\end{pmatrix}$ 2

   $\begin{pmatrix}a\\ b\end{pmatrix}=\dfrac{1}{30}\begin{pmatrix}1.35\\ 775.5\end{pmatrix}=\begin{pmatrix}0.045\\ 25.85\end{pmatrix}$ 4

   $a=0.045$, $b=25.85$ 1

4. (a) (i) $\dfrac{d}{dx}\left(\dfrac{x^4-4x^3+4x^2}{x^2}\right)=\dfrac{d}{dx}(x^2-4x+4)$

   $=2x-4$ 2

   (ii) $\dfrac{d}{dx}(3\sin x-\cos x)=3\cos x+\sin x$ 2

   (iii) $\dfrac{d}{dx}(3e^{4x})=12e^{4x}$ 2

   (b) Surface area of box $=x^2+4hx$

   $h=\dfrac{20-x^2}{4x}$ 2

   Volume $V=hx^2=x^2\left(\dfrac{20-x^2}{4x}\right)$ 2

   $V=\dfrac{20x^2}{4x}-\dfrac{x^4}{4x}=5x-\tfrac{1}{4}x^3$ 2

   $\dfrac{dV}{dx}=5-\tfrac{3}{4}x^2$, $\dfrac{d^2V}{dx^2}=-\tfrac{3}{2}x$ 2

   $5-\tfrac{3}{4}x^2=0$, $x^2=\dfrac{20}{3}$ 2

   $x=\pm2.582$ 1

   Maximum value is when $x=2.582$,
   $\dfrac{d^2y}{dx^2}$ is $-$ ve. 1

Marks   **Marks**

$$h = \frac{20 - 2.582^2}{4 \times 2.582} = 1.29 \text{ cm} \qquad 2$$

(b) Area $= \displaystyle\int_1^3 (x^2 + 6)dx = \left[\frac{x^3}{3} + 6x\right]_1^3 \qquad 4$

$$= (9 + 18) - (\tfrac{1}{3} + 6) \qquad 2$$

$$= 20\tfrac{2}{3} \qquad 2$$

5. (a) (i) $\displaystyle\int x^2\, dx = \frac{x^3}{3} + c \qquad 2$

6. Angle between the vectors is $\dfrac{\pi}{6}$ or $30^\circ \qquad 1$

(ii) $\displaystyle\int (x^3 + 6x)dx = \frac{x^4}{4} + \frac{6x^2}{2} + c$

$$V_R^2 = 16^2 + 24^2 - 2 \times 16 \times 24 \times \cos 150^\circ \qquad 3$$

$$= 256 + 576 + 768 \cos 30^\circ \qquad 2$$

$$= \frac{x^4}{4} + 3x^2 + c \qquad 4$$

$$= 1497 \qquad 2$$

$$V_R = \sqrt{1497} = 38.7 \qquad 2$$

(iii) $\displaystyle\int_1^4 x^{\frac{1}{2}} dx = \left[\frac{x^{\frac{3}{2}}}{\frac{3}{2}}\right]_1^4 \qquad 3$

$$\frac{25}{\sin \alpha} = \frac{38.7}{\sin 150^\circ} \qquad 3$$

$$= \left(\frac{4^{\frac{3}{2}}}{\frac{3}{2}}\right) - \left(\frac{1^{\frac{3}{2}}}{\frac{3}{2}}\right) \qquad 1$$

$$\sin \alpha = \frac{25 \times \sin 150^\circ}{38.7} = 0.3230 \qquad 3$$

$$\alpha = 18.84^\circ \ (0.329 \text{ rad}) \qquad 2$$

$$= \frac{16}{3} - \frac{2}{3} = \frac{14}{3} = 4\tfrac{2}{3} \qquad 2$$

Resultant is $38.7 \sin (\theta + 0.329)$ volts $\qquad 2$

Index

Books in the Granada TEC Series

MATHEMATICS FOR TECHNICIANS
Level 1
D. J. Hancox
0 246 11537 8

MATHEMATICS FOR TECHNICIANS
Level 2
D. J. Hancox
0 246 11725 7

MATHEMATICS FOR TECHNICIANS
Level 3
D. J. Hancox
0 246 11934 9

ELECTRICAL AND ELECTRONIC PRINCIPLES
Level 2
Rhys Lewis
0 246 11575 0

ELECTRICAL AND ELECTRONIC PRINCIPLES
Level 3
Rhys Lewis
0 246 11814 8   In preparation

ELECTRONICS FOR TECHNICIANS
Level 3
A. V. Smith
0 246 11488 6

ELECTRICAL AND ELECTRONIC APPLICATIONS
Level 2
A. V. Smith
0 246 11609 9   In preparation

DIGITAL TECHNIQUES FOR TECHNICIANS
Levels 2 and 3
J. McAllister
0 246 11787 7   In preparation